If And Only If In Analysis

Robert R. Dobbins
Fordham University

University Press
of America™

Copyright © 1977 by

University Press of America™

division of
R.F. Publishing, Inc.
4710 Auth Place, S.E., Washington, D.C. 20023

0–8191–0344–6

To My Mother
   and Father

PREFACE

The original idea for this book took its rise from the excellent, somewhat unique, book of Gelbaum and Olmsted, <u>Counterexamples in Analysis</u>. This fine book has served undergraduate mathematics majors and graduate students quite well for over ten years. The thought occurred that a similar book, aimed mainly at freshman and sophomore mathematics majors, might also be found to be a valuable contribution. A number of colleagues encouraged this thought and spoke favorably of such a project.

At the start, my plan was to model this book closely upon the format of Gelbaum and Olmsted. However, the content would differ. It would be a book treating the calculus of one and several variables. It was inevitable that there would be some overlap in topics, but it was my purpose to avoid most areas that a student would not encounter during the first two years of analysis.

In the course of writing this book, my original plan took on a number of modifications. The first was that I deemed it wiser to restrict myself to the calculus of a single variable. The reason for this was twofold. I did not wish the book to become unduly lengthy. Secondly, I wished to learn the type of reception such a book would have. There would be time in the future to write a similar book for the calculus of several variables. Again, while writing this book, I came to the judgement that the exact format of Gelbaum and Olmsted's book might not be the most appropriate one for the beginning student in mathematics. Hence, in the evolution of writing the book, the form came to be a type of commentary on the major theorems in the calculus of one variable.

However, my chief purpose in writing the book remained steadfast, namely to enable the student to dig in and to see precisely what a theorem says and what it does not say. When are the conditions necessary for a given conclusion? When are they sufficient for the conclusion but not required? Can you show this by apposite examples? I thought that the title "If and Only If in Analysis" would express my purpose as well as any other.

This book is not intended to be a text although it will be, it is hoped, a valuable supplement to most standard texts. In the first place, there are no proofs for the stated theorems. Their meaning is commented upon by way of examples or counterexamples. The proofs of most theorems can be found in the standard texts. Again, there are no exercises in this book. But although I have tried to write this book in an easy and readable style, it cannot be read through like a novel. Some of the examples that emphasize the point of a theorem do so immediately. Others, however, are cited but require some work on the part of the reader. For instance, if a function is said to be continuous at a single point and at no other point, it is not immediately obvious that such a function exists.

Not all the examples that are used to illustrate a theorem are simple, but I have attempted to avoid esoteric examples as far as possible. If a single example would illustrate the principal point of a number of theorems, I would use it rather than construct many such examples. The reason is that while I would wish the reader to have a catalog of such examples, yet I would hardly desire to make the catalog too large.

The viewpoint of this book is theoretical. None of the techniques of the calculus, differentiation or integration are considered. These techniques are assumed to be known. The elementary functions and their properties are likewise assumed to be known by the reader.

I would hope that this book and its examples would prove to be useful to the instructor of any calculus course. But the book has been written chiefly for the student. By studying appropriate examples and counterexamples in the context of otherwise dry theorems, I would wish that the student comes to obtain a deeper insight into the meaning of the theorems and comes to know just what if and only if means.

My first expression of appreciation must go to my students who have heard me speak on this material over the last several years and who have been a true source of inspiration to me. But even more than to these, I am deeply grateful to Mrs. Catherine DeRosa who has typed this manuscript for me with great skill, patience, and kindness.

Robert R. Dobbins
Fordham University
August 8, 1977

TABLE OF CONTENTS

I. INTRODUCTION. SETS AND FUNCTIONS.

1. A knowledge of set theory is essential in virtually every branch of modern mathematics, and it should not be assumed that set theory consists merely of the elementary relations every student has come to know from introductory courses in mathematics. Set theory is a broad topic and many of its theorems are rather profound. This said, however, the set theoretic background required for this book is relatively elementary and we will presume the reader's general acquaintance with it.

We may let our intuitive notion serve as our definition of a set. A set is the collection of all objects that satisfy some given conditions -- the brown cows in Delaware, the redwood trees in California, the prime numbers, the positive rational numbers, etc.

The set consisting of all real numbers which are unequal to themselves, that is the collection of real numbers x such that $x \neq x$, is also a set. We call this set the empty or null set $\emptyset$ since there are no real numbers with this property.

The properties that determine whether or not an object belongs to a given set must, of course, be well defined. However, this does not imply that we are always able to determine a given object's membership in a set or the lack thereof. For example, the set of real numbers with only a finite number of threes in the decimal expansion is a well defined set. But it is not known whether or not the base of the natural logarithms belongs to this set.

Given two sets A and B, the set A is a subset of the set B if every element belonging to A is also an element of the set B. This is written as A ⊂ B or B ⊃ A. If every element of A is an element of B and if there is an element of the set B which does not belong to the set A, then A is called a proper subset of B. Two sets are equal, that is A=B, if each set has precisely the same elements. Equality is proven by showing A ⊂ B and B ⊂ A. If A is not a subset of B, then there must be some element in A which does not belong to B. For this reason the empty set, ∅, which has no elements, is a subset of every set.

When we come to discuss functions, we will have to use a certain number of set theoretical operations. In order to fix the notation, we shall briefly review these operations.

The union of two sets A and B, A ∪ B, is defined

to be the set of elements that belong to A or to B or to both sets. $A \cup B = \{x: x\epsilon A,$ or $x\epsilon B,$ or $x\epsilon A$ and $x\epsilon B\}$. (We will follow the usual curly bracket notation for a set, $\{x: x$ has the property P$\}$.)

The intersection of two sets A and B, $A \cap B$, is defined to be the set of elements that belong both to A and to B. $A \cap B = \{x: x\epsilon A$ and $x\epsilon B\}$. If two sets have no elements in common, they are called disjoint and $A \cap B = \emptyset$. These definitions can easily be extended to unions and intersections of families of sets.

There are two additional set operations we will need, the difference of two sets and complementation. The difference of two sets A and B, $A - B$, is defined to be the set of elements that belong to A but do not belong to B. $A - B = \{x: x\epsilon A$ and $x \notin B\}$.

In order to define the complement of a set, however, a bit more care must be taken since complement is a relative notion. Let X be the set in question and let A be a subset of X, $A \subset X$. The complement of A in X, denoted by $C_X A$, consists of those elements which belong to X but do not belong to A. $C_X A = \{x: x\epsilon X$ and $x \notin A\}$. It may be that A is also a subset of another set Y. We may then take the com-

3.

plement of A in Y, $C_Y A$, but, in general, this is not the same set as the set $C_X A$.

For example, Let $X = \{0,1,2,3,4\}$ and let $A = \{1,2,3\}$. Then $C_X A = \{0,4\}$. Now if $Y = \{1,2,3,4,5\}$, A is also a subset of Y, but in this case $C_Y A = \{4,5\}$.

When it is clear from the context what the set X with respect to which we take the complement is, we shall omit the subscript. In particular, $CX = \emptyset$ and $C\emptyset = X$. But our main point should be clear -- complement is a relative notion.

We might note at this point that the operation of complementation is defined only in the case where one set is contained in the other, whereas the difference operation has no such requirement. Nevertheless, the two operations are strikingly similar, and with good reason. Let A and B both be subsets of the set X. Then $A - B = A \cap C_X B$.

Two sets A and B are said to have the same cardinal number if they can be put into one-to-one correspondence with each other. In particular, two finite sets have the same cardinality if each of the two sets has the same number of elements. For instance, let $A = \{a,b,c,d\}$ and let $B = \{1,5,7,9\}$. Then the set A and the set B have the same cardinality.

4.

If a set A is not finite, then it is infinite.
One way to characterize an infinite set is as follows:
an infinite set can be put into one-to-one correspondence with a proper subset of itself. For example,
the set of positive integers can be put into one-to-one correspondence with the set of positive even
integers which is a proper subset of it. This is, of
course, impossible for finite sets.

A set is countably infinite or denumerable if it
can be put into one-to-one correspondence with the
positive integers. For example, the set of rational
numbers is countably infinite. In this book we shall
use the term countable to denote a set that is either
countably infinite or finite. A set that is not
countable is uncountable. The set of real numbers is
an example of an uncountable set. It would be well
for the reader to show that the set $\{0 \leq x \leq 1\}$ is an
uncountable set by showing that it cannot be put into
one-to-one correspondence with the set of positive
integers.

Another set theoretical notion for which we shall
have some use is that of the Cartesian product of two
sets. Let A and B be two given sets. The Cartesian
product of A and B, AxB, is the set $\{(a,b):a\epsilon A \text{ and } b\epsilon B\}$.

5.

Hence the Cartesian product is a set consisting of ordered pairs (a,b) such that the first element is a member of A and the second element a member of B. Two ordered pairs, say (a,b) and (c,d), are equal if and only if a=c and b=d. The Cartesian product of A and B is empty, that is AXB = $\emptyset$, if and only if A = $\emptyset$ or B = $\emptyset$ (or, of course, both A and B could be empty.).

The further elementary set properties are easily available in any suitable text and we will assume the reader is familiar with them.

2.   Functions

Throughout this book we shall be considering functions or maps -- the terms will be used interchangeably -- on the real numbers or on subsets of the real numbers. But since the notion of function is of broader scope and since it will not add any appreciable difficulty to our discussion, it seems better to define function in a context of greater generality, that is in terms of the more primitive notion of sets.

By a function or map f: X$\rightarrow$Y on a set X into a set Y we mean a rule that assigns each element in X a unique element, generally denoted by f(x), in Y.

Hence we can look upon a function as a subset of the Cartesian product of X and Y, XxY, that is $f \subset XxY$. However, some care is necessary since not every subset of XxY represents a function. For example, consider the set of ordered pairs $\{(a,\alpha), (b,\beta), (c,\gamma) (a,\delta)\}$. The elements b and c in X are assigned the elements β and γ in Y. But what element in Y is assigned to a? This set of ordered pairs does not represent a function. Or to take what may be a more familiar example, consider the set of ordered pairs $\{(x,y); x^2 + y^2 = 1\}$. This set does not represent a function. However the sets $\{(x,y):x^2 + y^2 = 1, y > 0\}$ and $\{(x,y):x^2 + y^2 = 1, y < 0\}$ do represent functions. Hence our definition: a function is a collection of ordered pairs such that no two distinct pairs have the same first element.

Any number of pairs can have the same second element. In fact every pair in the collection may have the same second element. In this case we have the constant function. But if f is to be a function and if (a,α) and (a,δ) are members of f, then $\alpha = \delta$.

The set X is called the domain of f, $D(f)$, that is the domain of f is the collection of all x for which there is a unique y such that $(x,y)\epsilon f$. Given

7.

any subset $A \subset X$, the image of A under f, $f(A)$, is the set of elements in Y such that $y = f(x)$ for some x in X: $f(A) = \{y : \exists x, x \in A \text{ and } y = f(x)\}$. The image of X under f, $f(X)$, is called the range of f, $R(f)$. In general, $f(X)$ will be a proper subset of Y. However, if $f(X) = Y$, the function f is said to be onto Y or surjective.

A function $f: X \rightarrow Y$ is said to be one-to-one or injective if $f(x) = f(y)$ implies $x = y$ (or $x \neq y$ implies $f(x) \neq f(y)$). Functions which are one-to-one from X onto Y, that is, functions which are both surjective and injective, are called bijective.

We have already defined the image under f of a subset of X. We now define the inverse image. Let B be some subset of Y. The inverse image of $B, f^{-1}(B)$, is a subset of X, namely those $x \in X$ for which $f(x) \in B$; $f^{-1}(B) = \{x : f(x) \in B\}$. Because it is more well behaved under set operations, as we shall see below, the inverse image of a set is often of greate interest than the image.

In general, f^{-1} is not to be thought of as a function. It usually is not! However, in the event that f is both one-to-one and onto, there does exist a function, call it g, such that $g(f(x)) = x$ and

8.

$f(g(y)) = y$ for all x and y. The function g is called the inverse of f and, for obvious reasons, it is commonly denoted by f^{-1}.

3. Functions and Set Operations.

We now wish to consider the behavior of images and inverse images under set operations. The reader should attempt to prove each of the relations given below. We will merely state the results and give some concrete examples to show that equality need not hold in each case.

Let $f: X \to Y$ where $A_\alpha \subset X$ and $B_\alpha \subset Y$.

$$f(A_1 \cup A_2) = f(A_1) \cup f(A_2) \qquad (1)$$

$$f(\bigcup_\alpha A_\alpha) = \bigcup_\alpha f(A_\alpha) \qquad (1')$$

$$f(A_1 \cap A_2) \subset f(A_1) \cap f(A_2) \qquad (2)$$

$$f(\bigcap_\alpha A_\alpha) \subset \bigcap_\alpha f(A_\alpha) \qquad (2')$$

$$A_1 \subset A_2 \;\to\; f(A_1) \subset f(A_2) \qquad (3)$$

$$f(A_1) - f(A_2) \subset f(A_1 - A_2) \qquad (4)$$

$$f(A) = \emptyset \;\to\; A = \emptyset \qquad (5)$$

$$f^{-1}(\bigcap_\alpha B_\alpha) = \bigcap_\alpha f^{-1}(B_\alpha) \qquad (6)$$

$$f^{-1}(\bigcup_\alpha B_\alpha) = \bigcup_\alpha f^{-1}(B_\alpha) \qquad (7)$$

$$f^{-1}(B_1 - B_2) = f^{-1}(B_1) - f^{-1}(B_2) \qquad (8)$$

$$B_1 \subset B_2 \;\to\; f^{-1}(B_1) \subset f^{-1}(B_2) \qquad (9)$$

9.

$$f^{-1}(B) = \emptyset \leftrightarrow B \cap f(X) = \emptyset \qquad (10)$$

$$A \subset f^{-1}(f(A)) \qquad (11)$$

$$f(f^{-1}(B)) \subset B \qquad (12)$$

A comparison of (6) with (2) and of (8) with (4) shows that the inverse image is better behaved than the image under set operations. Two simple examples can serve to illustrate the strict inclusion in (2) and (4). The reader may amuse himself by constructing more elaborate examples.

ad 2 a). Let $f: \{a,b,c,d\} \nrightarrow \{\alpha,\beta,\gamma\}$ be defined by $f(a) = \beta$; $f(b) = \gamma$; $f(c) = \alpha$, $f(d) = \beta$. Take $A_1 = \{a,$ $A_2 = \{b,d\}$ so that $A_1 \cap A_2 = \{b\}$. Then $f(A_1 \cap A_2) = \{\gamma\}$ $f(A_1) = \{\beta,\gamma\}$, $f(A_2) = \{\beta,\gamma\}$, and so $f(A_1) \cap f(A_2) = \{\beta,\gamma\}$ Hence $f(A_1 \cap A_2) \subset f(A_1) \cap f(A_2)$.

ad 2 b). Let $f:X \rightarrow Y$ where $X = [-1,1]$ and $Y = \{y: y = (1-x^2)^{\frac{1}{2}}\} = [0,1]$. Take $A_1 = [-1,0]$ and $A_2 = [0,1]$, so that $A_1 \cap A_2 = \{0\}$. Then $f(A_1 \cap A_2) = f(0)$ $= 1$ and $f(A_1) = f(A_2) = [0,1]$. Hence $f(A_1 \cap A_2) \subset f(A_1)$ $\cap f(A_2)$ properly.

The same two examples can be used to illustrate (4).

10.

ad 4 a). $f(A_1)-f(A_2) = \emptyset$ and $f(A_1-A_2)= f(\{a\})= \beta$,
so that $f(A_1)-f(A_2) \subset f(A_1-A_2)$.

ad 4 b). $f(A_1)-f(A_2) = \emptyset$ and $f(A_1-A_2)= f([-1,0))= [0,1]$, so that $f(A_1)-f(A_2) \subset f(A_1-A_2)$.

It should be noted that in both of these examples f is not one-to-one. In fact, it can be shown that equality holds in (2) and (4) if and only if f is injective.

Now consider (11) and (12). For (11) the same functions as above will illustrate the proper in-clusion.

ad 11 a). Take $A = \{a,b\}$. Then $f(\{a,b\})=\{\beta,\gamma\}$, and $f^{-1}(f(\{a,b\})) = f^{-1}(\{\beta,\gamma\}) = \{a,b,d\}$. Hence, $A \subset f^{-1}(f(A))$ properly.

ad 11 b). Take $A = [0,1]$. Then $f([0,1])= [0,1]$, and $f^{-1}(f([0,1])) = f^{-1}([0,1]) = [-1,1]$. So that again $A \subset f^{-1}(f(A))$ properly.

As was the case with (2) and (4), we have equality in (11) if and only if f is injective.

In order to illustrate (12) consider the follow-ing function. Let $f:X \rightarrow Y$ where $X = [-1,1]$ and $Y = \{1,0,-1\}$, and

11.

$$f(x) = \begin{cases} 1 & 0 < x \leq 1 \\ 0 & x = 0 \\ 1 & -1 \leq x < 0 \end{cases}$$

Note that we have arranged it so that f is not onto
Y. Let $B = Y$. Then $f^{-1}(B) = f^{-1}(Y) = [-1,1]$ and
$f(f^{-1}(B)) = f([-1,1]) = \{0,1\} \subset B = \{1,0,-1\}$.
In contrast with (11), we get equality in this case,
that is, $f(f^{-1}(B)) = B$, if and only if f is surjective.
However, it can be shown -- and this result is true
in general -- that $f(f^{-1}(B)) = B \cap f(X)$.

There remains one more set operation to consider,
namely complementation. Again, let $f:X \rightarrow Y$. As above,
we find that the inverse image of a set is well behaved.

$$f^{-1}(CB) = Cf^{-1}(B) \tag{13}$$

But the case of the image of a set under complementation is slightly more involved. The results are best
expressed in the form of two lemmas which we leave to
the reader to prove.

Lemma I. The function f is injective if and

only if $f(CA) \subset Cf(A)$. (14)

Lemma II. The function f is surjective if and

only if $Cf(A) \subset f(CA)$ (15)

It should be noted that these two lemmas together imply that f(CA) = Cf(A) if and only if f is bijective.

Strict inclusion is easily demonstrated in each case. For (14) we can take any injective function, say $f:X \to Y$ where $X = \{a,b,c,d\}$ and $Y = \{\alpha, \beta, \gamma, \delta, \varepsilon\}$. Let $f(a) = \alpha$, $f(b) = \beta$, $f(c) = \gamma$, and $f(d) = \delta$. Now take $A = \{a,b\}$. Then $f(CA) = f(\{c,d\}) = \{\gamma, \delta\}$ and $Cf(A) = Cf(\{a,b\}) = \{\gamma, \delta, \varepsilon\}$. Hence we see that the inclusion is proper.

On the other hand, to illustrate (15) we need only to take any surjective function. Consider $f:X \to Y$ where $X = [-1,1]$ and $Y = \{0,1\}$, and

$$f(x) = \begin{cases} 1 & 0 < x \leq 1 \\ 0 & x = 0 \\ 1 & -1 \leq x < 0 \end{cases}$$

Take $A = [0,1]$. Then $Cf(A) = C\{0,1\} = \emptyset$ and $f(CA) = f([-1,0)) = \{1\}$. Hence we again obtain strict inclusion.

4. Elementary Operations with Functions.

It is possible to combine functions to get new functions by using the elementary arithmetic operations.

13.

i. $(f+g)(x) = f(x) + g(x)$.

ii. $(f-g)(x) = f(x) - g(x)$.

iii. $(\alpha f)(x) = \alpha f(x)$.

iv. $(f \cdot g)(x) = f(x) \cdot g(x)$.

v. $(f/g)(x) = f(x)/g(x)$.

However, some care must be taken with the domain of definition of these functions. If we denote the domains of f and g by $D(f)$ and $D(g)$ respectively, the domain of $(f+g)$, $(f-g)$, and $f \cdot g$ is $D(f) \cap D(g)$, that is the set common to the domains of both f and g. The domain of f/g is $D(f) \cap D(g) - \{x: g(x) = 0\}$; and the domain of αf is clearly the same as the domai of f.

For example if $f(x) = (x-2)^{\frac{1}{2}}$ and $g(x) = (1-x)^{\frac{1}{2}}$, then the sum, product, difference, and quotient of f and g would not be defined for any real number since the set which comprises the common domain of f and g is null set.

Given two functions f and g, we define the composite function f∘g (read f circle g) as $(f \circ g)(x) = f(g(x))$. The composite function must be carefully distinguished from the product of two functions. For instance, whereas $f \cdot g = g \cdot f$, it is, in general, not

true that f∘g = g∘f. The domain of f∘g, D(f∘g), is
$\{x \varepsilon D(g): g(x) \varepsilon D(f)\}$. We might also note that
(g+h)∘f = g∘f + h∘f but that for arbitrary functions
f∘(g+h) ≠ f∘g + f∘h.

In order to illustrate the noncommutivity of
composite functions as well as the nondistributivity
of composition on the left, consider the following
example. Let $f(x) = \sin x$, $g(x) = x^2$, and $h(x) = x^2$.
All three functions are defined on the entire real
line. Then we have

$$[f \circ (g+h)](x) = f(2x^2) = \sin(2x^2).$$
$$(f \circ g)(x) = (f \circ h)(x) = \sin(x^2), \text{ and}$$
$$(g \circ f)(x) = (h \circ f)(x) = \sin^2 x.$$

Therefore, $[(g+h) \circ f](x) = (g+h)(\sin x) = 2 \sin^2 x \neq$
$[f \circ (g+h)](x)$ and $[f \circ (g+h)](x) \neq (f \circ g)(x) + (f \circ h)(x)$.

As an added indication of the care we must take
with order in composition, suppose $f < g$, that is,
$f(x) < g(x)$ for all x. It is not true that h∘f < h∘g.
To see this we may simply take $h(x) = -x$. On the
other hand, it is quite evident that f∘h < g∘h for
all h for which both sides of the inequality are
defined.

We close this chapter by noting that if
f∘g: X→Y is a one-to-one onto mapping, so that the

15.

inverse function is defined, $(f \circ g)^{-1}$ is not $f^{-1} \circ g^{-1}$. Rather the inverse of the composite function is given by $(f \circ g)^{-1} = g^{-1} \circ f^{-1}$.

II.   LIMITS AND CONTINUITY.

1.   Limits.

The concept of limit is fundamental to the calculus but, unfortunately, the limit idea is one with which the beginning student often has the greatest problems. What is meant by the statement $\lim_{x \to a} f(x) = A$? In a vague and intuitive way, the idea is roughly this: the function f takes on a value as close to A as you like as long as x is close enough to a. We want to note two things.

1.   x is to be close to a, but nothing is said about x being equal to a. In fact, we want x to be unequal to a.

2.   We do not care about f(a), the value of f at a. The function can be defined or not at x = a. It does not matter! And even if the function is defined at x = a, its value need not be A in order that the limit of f as x approaches a be A.

Consider the function $f(x) = x \sin(1/x)$ where the function is defined for all values of x except x = 0. We don't say anything about the value of f

17.

at x = 0. Now look at how this function behaves as x gets close to zero. The sine can take on only values between plus and minus one. Therefore, we have

$$|x \sin(1/x)| = |x| \; |\sin(1/x)| \leq |x|,$$

and we see that f(x) is as close to zero as we wish as long as x is close enough to zero. In this case we write

$$\lim_{x \to 0} \; x \sin(1/x) = 0.$$

In line with what was said above, we note that x is close to zero but x ≠ 0. This has to be so here since this function is not defined at x = 0. But we could define it there. A sensible student would say let us define f(0) = 0. A more perverse student might say let us define f(0) = 10.3. Whether defined by the sensible or perverse student, however, the fact remains that

$$\lim_{x \to 0} \; x \sin(1/x) = 0$$

because, as we said above, the limit is concerned with how the function behaves near x = 0, not at x = 0!

Now the words "near" and "close" are fine for intuition but not very mathematical. We need a rigorous definition of the statement $\lim_{x \to a} f(x) = A$.

Definition.

Lim $f(x) = A$ if given any $\varepsilon > 0$, there exists
$x \to a$
a $\delta > 0$ such that $|f(x) - A| < \varepsilon$ for all x such that $0 < |x-a| < \delta$.

Note what $0 < |x-a| < \delta$ means. It means $x \neq a$ as we emphasized above!

We can now apply this definition to our previous example where $A = 0$ and $a = 0$. Then, given $\varepsilon > 0$, what δ can we choose so that $|x \sin(1/x) - 0| < \varepsilon$? As we saw

$$|x \sin(1/x) - 0| = |x| \, |\sin(1/x)| \leq |x|,$$

and this expression is certainly less than ε if we take $\delta = \varepsilon$. In this example it was easy to choose δ as a function of ε so that the desired inequality held. In other cases it may not be as easy to find δ in terms of ε, but the idea of limit remains the same.

Of course the limit of a function at a point need not exist. If, rather than the function we had chosen in the illustration above, we had taken

19.

$f(x) = \sin(1/x)$, then $\lim\limits_{x \to 0} \sin(1/x)$ would not exist.

Now let us consider a function called the greatest integer function, $f(x) = [x]$, where $[x]$ is the greatest integer equal to or less than x. For example, $[-1.5] = -2, [\pi] = 3$, etc. What is $\lim\limits_{x \to 0} [x]$? If we take some small interval containing $x = 0$, say $-\frac{1}{4} < x < \frac{1}{4}$, and consider all the values of $[x]$ in this interval except its value at $x = 0$, it is clear that $[x] = -1$ if x is negative and $[x] = 0$ if x is positive in the interval. The same result holds no matter what small interval we take about $x = 0$. The greatest integer function behaves in a similar way at every integer -- the limit fails to exist. However, the limit does exist for every real number which is not an integer. A graph of this function wil easily convince the reader that this is so.

The greatest integer function naturally leads us to a new idea, the idea of one sided limits. Although the limit of the greatest integer function does not exist at the origin (or at any other integer if we approach the origin only from the left side or only from the right side, the limits do exist. But they are unequal! We have

$$\lim_{x \to 0^-} [x] = -1 \qquad \text{and} \qquad \lim_{x \to 0^+} [x] = 0,$$

here $\lim_{x \to 0^-}$ means the origin is approached from the

eft and $\lim_{x \to 0^+}$ means the origin is approached from

he right.   More generally, we can define the one-

sided limits of a function.

Definition.

1.   The limit of $f(x)$ as x approaches a from

the right is A -- written as $\lim_{x \to a^+} f(x) = A$ -- if

given any $\varepsilon > 0$, there exists a $\delta > 0$ such that

$|f(x) - A| < \varepsilon$ for all x such that $0 < x-a < \delta$.

2.   The limit of $f(x)$ as x approaches a from

the left is B -- written as $\lim_{x \to 0^-} f(x) = B$ -- if

given any $\varepsilon > 0$, there exists a $\delta > 0$ such that

$|f(x) - B| < \varepsilon$ for all x such that $0 < a-x < \delta$.

In order that a function have a limit at a

point, it is necessary and sufficient that the limits

both from the left and from the right exist and are

equal. We saw both left and right limits exist at the

origin for the greatest integer function but these

are unequal. Hence, $\lim_{x \to 0} [x]$ does not exist.

There are two more topics we should consider before we discuss the limits of sums, products, and quotients of functions, namely, limits at infinity and infinite limits. For example, as x gets larger and larger, it is clear that $1/x^2$ gets closer and closer to zero. We write this as $\lim_{x \to \infty} 1/x^2 = 0$.

Definition.

1. The limit of f as x increases without bound is A -- written as $\lim_{x \to \infty} f(x) = A$ -- if given any $\varepsilon > 0$ there exists some number $N > 0$ such that $|f(x) - A| < \varepsilon$ for all $x > N$.

2. The limit of f as x decreases without bound is B -- written as $\lim_{x \to -\infty} f(x) = B$ -- if given any $\varepsilon > 0$, there exists some number $N < 0$ such that $|f(x) - B| < \varepsilon$ for all $x < N$.

Of course, the limits at infinity need not exist. For instance, $\lim_{x \to \infty} \sin x$ does not exist.

Finally, as x approaches a (or grows or decreases without bounds), the value of the function may grow or decrease without bounds. Hence we make the following definition.

22.

Definition.

1. The function f increases without bound as x approaches a -- written as $\lim_{f \to a} f(x) = \infty$ -- if given any number $N > 0$, we can find a $\delta > 0$ such that $f(x) > N$ for all such x such that $0 < |x-a| < \delta$.

2. The function f decreases without bound as x approaches a -- written as $\lim_{x \to a} f(x) = -\infty$ -- if given any number $N < 0$, we can find a $\delta > 0$ such that $f(x) < N$ for all x such that $0 < |x-a| < \delta$.

For example, let $f(x) = 1/(x-a)^2$. Then $\lim_{x \to a} 1/(x-a)^2 = \infty$. On the other hand, $\lim_{x \to a} 1/(x-a)$ does not exist, for as x approaches a from the right the values of the function grow without bound, whereas as x approaches a from the left the values decrease without bound,

Combining the last two definitions, we can consider such expressions as $\lim_{x \to \infty} f(x) = \infty$. For example, $\lim_{x \to \infty} x^{\frac{1}{2}} = \infty$. What is $\lim_{x \to -\infty} f(x)$ for this same function?

To take an example that requires more thought, let $f(x) = 1/x^2 \sin^2(\pi/x)$. What is $\lim_{x \to 0} f(x)$? You might be tempted to say immediately that f grows

23.

without bounds since $\sin^2(\pi/x)$ is certainly bounded and $\lim\limits_{x\to 0} 1/x^2 = \infty$. But the limit of this function at the origin does not exist. It is true that there are points as close to the origin as we wish where the function takes on values as large as we wish; for example, $x = 2/(2n+1)$ with n an integer. On the other hand, the points $x = 1/n$ with n an integer are as close to the origin as we wish but the value of the function is zero at these points.

Theorem 1.

Let f and g be functions defined on some inter-val containing $x = a$ (it is not necessary that the functions be defined at $x = a$). Suppose that the limits $\lim\limits_{x\to a} f(x)$ and $\lim\limits_{x\to a} g(x)$ both exist. Then the limits of the sum $f(x) + g(x)$ and the product $f(x)\cdot g(x)$ both exist and are given by $\lim\limits_{x\to a} (f(x)+g(x))$

$= \lim\limits_{x\to a} f(x) + \lim\limits_{x\to a} g(x)$ and $\lim\limits_{x\to a} f(x)\cdot g(x)$

$= (\lim\limits_{x\to a} f(x)) (\lim\limits_{x\to a} g(x))$. Moreover, if $\lim\limits_{x\to a} g(x) \neq 0$, the limit of the quotient $f(x)/g(x)$ exists and is given by $\lim\limits_{x\to a} f(x)/g(x) = \lim\limits_{x\to a} f(x)/\lim\limits_{x\to a} g(x)$.

It should be noted that the limits of the sum, product, and quotient can all exist as x approaches a, even if neither $\lim\limits_{x \to a} f(x)$ nor $\lim\limits_{x \to a} g(x)$ exist. A single

example is sufficient to illustrate all three cases.

Consider the functions $f(x) = \begin{cases} 1 & x \text{ rational} \\ -1 & x \text{ irrational} \end{cases}$ and

$g(x) = \begin{cases} -1 & x \text{ rational} \\ 1 & x \text{ irrational} \end{cases}$. Both functions are well

defined on the entire real line and neither function

has a limit at any point. Nevertheless, their sum,

product, and quotient are the constants 0, -1, and

-1 respectively. Consequently, the limits of sum,

product, and quotient exist at every point.

A question now naturally arises. The theorem

gives a sufficient condition for the sum, product, and

quotient to have a limit, but the condition is not

necessary. What can we say if the limit of one

function exists at the point but the other function

does not have a limit at the point?

Consider the case in which the function f has a

limit at a given point but g does not have a limit at

the point. It is not difficult to see that $\lim\limits_{x \to a} (f(x)$

+ g(x)) cannot exist. For if it did, then

$\lim\limits_{x \to a} g(x) = \lim\limits_{x \to a} ((f+g)(x) - f(x))$ must also exist

and this is contrary to the given hypothesis.

Now look at the similar question with products

in place of sums. You might be tempted to say that

the result is the same, namely, the product does not have a limit. You might argue that, if it did, then $\lim_{x \to a} g(x) = \lim_{x \to a} (f \cdot g)(x)/f(x)$ would exist. But is this so? Certainly if $\lim_{x \to a} f(x) = 0$ this argument will not work and, in this case, we can construct function to show that the limit of the product does exist. For example, we saw that $\lim_{x \to 0} \sin(1/x)$ does not exist. Let us take $g(x) = \sin(1/x)$ and $f(x) = x$.   Then $\lim_{x \to 0} (f \cdot g)(x) = 0$, so that in this case the limit of the product does indeed exist even though only one of the factors has a limit at the origin.

Theorem 2.

If $\lim_{x \to a} f(x) = A$, then the limit of the absolute value of f, $|f|$, exists at x=a and $\lim_{x \to a} |f(x)| = |A|$.

The converse of **this** theorem, however, is false as a function we considered above, $f(x) = \begin{cases} 1 & x \text{ rational} \\ -1 & x \text{ irrational} \end{cases}$, clearly demonstrates.

Theorem 3.

Let f and g be functions whose limits exist at x = a and let $f(x) \le g(x)$ for all x such that $0 < |x-a| < \delta$ where δ is arbitrary. Then $\lim_{x \to a} f(x) \le \lim_{x \to a} g(x)$.

26.

Suppose the inequality is a strict inequality, that is $f(x) < g(x)$ for all $0 < |x-a| < \delta$. Is there strict inequality in the conclusion of the theorem? The answer is no, as a simple example with show. Let $f(x) = |x-a|/2$ and $g(x) = |x-a|$. Then for all x with $0 < |x-a| < \delta$, $f(x) < g(x)$, but $\lim_{x \to a} f(x) = \lim_{x \to a} g(x) = 0$.

2. Continuity.

The idea of continuity is closely connected with that of limit. In fact, the ε-δ definitions of limit and continuity appear to be almost identical. But there are significant differences! For a function to be continuous at a point, say $x = a$, we must have the following three conditions fulfilled.

1. The function must have a limit at the point, $\lim_{x \to a} f(x) = A$.

2. The function must be defined at the point, that is, $f(a)$ is well defined.

3. The limit of the function must equal the value of the function at $x = a$, $f(a) = A$.

Hence, we make the ε-δ definition of continuity.

Definition.

A function f is said to be continuous at x = a, if given ε > 0, there exists a δ > 0 such that |f(x) - f(a)| < ε for all x such that |x-a|<δ .

Note that we have written |x-a|< δ and not 0 < |x-a|< δ . The reason is that a function that is continuous at a point, unlike a function that merely has a limit at a point, must be defined at that point Hence, in the inequality we must also consider points where x = a. A function that is continuous at every point in its domain of definition is simply called continuous.

One of the more important properties of continuous functions is what might be called persistence of sign.

Theorem 4.

Let f be continuous at x = a, and suppose f(a)> Then there is a δ > 0 such that f(x)> 0 for all x such that |x-a|< δ . An analogous result holds if f(a) < 0.

This theorem is important in many theoretical arguments, but it also has practical consequences.

For instance, consider a function defined and continuous on $-1 \le x \le 1$, and suppose $f(x) = 0$ for rational x. The reader should be able to apply the theorem to demonstrate that $f(x) = 0$ for all points in the interval.

If a function is not continuous at a point, it is said to be discontinuous at the point.

It is often useful to distinguish between types of discontinuities. Let us suppose that the function f is defined on an open interval and is such that the limits from the left and the right, $\lim_{x \to a^-} f(x)$ and $\lim_{x \to a^+} f(x)$, both exist at $x = a$, but that the function is discontinuous at this point. This can occur in various ways. The function may be such that

$\lim_{x \to a^+} f(x) \ne \lim_{x \to a^-} f(x)$. Or it may be that

$\lim_{x \to a^+} f(x) = \lim_{x \to a^-} f(x) \ne f(a)$. In such cases f is said to have a simple discontinuity or a discontinuity of the first kind at $x = a$. Otherwise the discontinuity is called an essential discontinuity or a discontinuity of the second kind.

An example of the first kind of discontinuity would be $f(x) = \begin{cases} 1 & x \ge a \\ -1 & x < a \end{cases}$. An example of the second kind of discontinuity would be $f(x) = \begin{cases} 1 & x \text{ rational} \\ -1 & x \text{ irrational} \end{cases}$.

We should also distinguish a special type of discontinuities of the first kind. Suppose the function is such that $\lim_{x \to a} f(x)$ exists, that is both left and right handed limits exist and are equal. Further, suppose that either the function is not defined at $x = a$ or, if $f(a)$ is defined, $\lim_{x \to a} f(x) \neq f(a)$. For example, consider the functions

$f(x) = x \sin(1/x)$ for $x \neq 0$ and $g(x) = \begin{cases} x \sin(1/x) & x \neq 0 \\ 1 & x = 0 \end{cases}$

Both f and g have discontinuities of the first kind at the origin. Moreover, $\lim_{x \to 0^+} f(x) = \lim_{x \to 0^-} f(x) = \lim_{x \to 0^+} g(x)$

$= \lim_{x \to 0^-} g(x) = 0$. Now if we define f at the origin to be $f(0) = 0$, and if we redefine g at the origin to be $g(0) = 0$, it is clear that f and g both become continuous at $x = 0$. Simple discontinuities of this type are called, for obvious reasons, removable discontinuities.

There is a vast variety to the discontinuities that a function may have. One function we have seen a number of times, $f(x) = \begin{cases} 1 & x \quad \text{rational} \\ -1 & x \quad \text{irrational} \end{cases}$, is discontinuous at every point, The function,

$f(x) = \begin{cases} x & x \text{ rational} \\ 0 & x \text{ irrational} \end{cases}$, is continuous at the origi

but discontinuous at every other point. Another interesting example is the function

$$f(x) = \begin{cases} 0 & x \text{ irrational} \\ 1/q & x = p/q \text{ in lowest terms} \end{cases}$$ (where p and q

are integers). This function is continuous at every

irrational number and discontinuous at every rational

number. It would be a profitable exercise for the

reader to verify the stated properties of these last

two functions.

We should mention that the last example above

is not a counterexample to the statement that two

continuous functions that agree on a dense set of

points are identical. (A set of real numbers is

called dense if every open interval contains a

point of this set. An example would be the set of

rational numbers.) Let $g(x) = 0$ for all x. Then g

is continuous and agrees with the function f in the

last example above on a dense set, namely the irra-

tionals. But these functions are not identical since

f is not continuous.

In the same way as we considered one-sided

limits, we can define continuity on the left or on

the right. For example, the function $f(x) = x^{\frac{1}{2}}$ is

continuous on the interval $(0, \infty)$. But since $\lim_{x \to 0} f(x)$

does not exist, this function is not continuous at

the origin. However, f is continuous from the right
at x = 0, that is $\lim_{x \to 0^+} f(x) = f(0)$.

Another example would be the greatest integer
function, $f(x) = [x]$. It is easy to see that, although discontinuous (simple discontinuities) at
every integer, this function is continuous from the
right at every integer.

The perceptive reader may have noticed that, in
our definition of continuity, we specified that the
function was to be defined on an open interval, say
(a,b). What of the continuity of a function on a
closed interval [a,b], or on a half open half closed
interval (a,b]? The situation is quite simply taken
care of. For f to be defined and continuous on [a,b],
we demand that f be continuous on (a,b) and continuous
from the right at x = a and from the left at x = b.
We deal with the intervals (a,b] and [a,b) in an
analogous fashion.

We should make the remark here that a function f,
continuous on a closed interval [a,b], can be extended
to a function h, continuous on the entire real line.
Thus, $h(x)\Big|_{[a,b]} = f(x)$, that is, f may be considered
to be the restriction of h to the interval [a,b]. To
see that our statement is well founded, we need only

observe that, since f is continuous [a,b], $\lim_{x \to a^+} f(x)$

and $\lim_{x \to b^-} f(x)$ both exist . Hence, if we define

$$h(x) = \begin{cases} \lim_{x \to a^+} f(x) & x \leq a \\ f(x) & a < x < b \\ \lim_{x \to b^-} f(x) & x \geq b \end{cases}$$

then h is the desired extension of f. But it is to be noted that f and h are different functions since their domains differ.

However, for the above argument to be applicable, it is essential that the interval on which the function is continuous be closed. For instance the function $f(x) = \sin(1/x)$ is continuous on $(0,1]$, but this function cannot be extended to a function continuous on the entire real line.

Theorem 1 considered the sum, product, and quotient of limits of functions. There is a theorem for continuous functions similar to this theorem. The sum, product, and quotient of continuous functions are continuous. But as with limits, the continuity of the sum, product, and quotient allows us to infer nothing about the continuity of the constituent functions. Again, if only one of the two functions is continuous, we can conclude results analogous to those found with limits about the con-

33.

tinuity of the sum and product.

Theorem 5.

If f is continuous at x = a, then |f| is con-
tinuous at that point.

The theorem gives a sufficient condition for
the continuity of |f|. In order to see that the
condition is not necessary, we need merely consider
the function

$$f(x) = \begin{cases} 1 & x \text{ rational} \\ -1 & x \text{ irrational} \end{cases}$$

When we considered the limits of sums, products,
and quotients, we neglected to consider the limits of
composite functions. This was not unintentional. We
have the following theorem.

Theorem 6.

If $\lim_{x \to a} g(x) = b$ and if f is continuous at b,
then $\lim_{x \to a}(f \circ g)(x) = f(b)$.

Note that the conclusion simply states
$\lim_{x \to a} f(g(x)) = f(\lim_{x \to a} g(x))$. What must be emphasised is
the continuity of f. If f fails to be continuous,
the conclusion may well be false as the following
example shows.

Let $f(x) = \begin{cases} 1 & x \geq 0 \\ -1 & x < 0 \end{cases}$ and let $g(x) = x$. Then

$f(\lim_{x \to 0} g(x)) = f(0) = 1$. But $\lim_{x \to 0} f(g(x)) = \lim_{x \to 0} f(x)$

fails to exist.

Of course, the composite function of two continuous functions is, as we would expect, continuous. But the composite function can be continuous without either of the constitutent functions being continuous. For consider the functions $f(x) = \begin{cases} 1 & x \text{ rational} \\ -1 & x \text{ irrational} \end{cases}$

and $g(x) = \begin{cases} x & x \text{ rational} \\ 0 & x \text{ irrational} \end{cases}$. Then $(f \circ g)(x) = 1$

and so $f \circ g$ is continuous for all x. Yet f is discontinuous for all x and g is discontinuous except at the origin. (Even more simply, we could have taken $f \circ f$ to illustrate this point.)

3. Uniform Continuity.

We have defined continuity at a point. If a function is continuous at every point of an interval, we say that the function is continuous on the interval. We now wish to consider continuity on intervals in somewhat greater detail.

If a function is continuous at each point where it is defined, then, as we have seen, for any $\varepsilon > 0$, we can find a $\delta > 0$ such that $|f(x) - f(x_o)| < \varepsilon$

for all x such that $|x-x_0| < \delta$. It is natural to think that if we choose a smaller ϵ, a smaller value of δ will be required. This is indeed so; δ is a function of ϵ. Now suppose f is continuous at x_0 and at x_1. Take $\epsilon = 1$. Then there is a δ, call it $\delta(1)$, such that $|f(x) - f(x_0)| < 1$ for all x such that $|x-x_0| < \delta(1)$. Since f is also continuous at x_1, does it follow that $|f(x) - f(x_1)| < 1$ for all x such that $|x-x_1| < \delta(1)$? The answer is, in general, negative. There is some δ such that $|x-x_1| < \delta$ implies $|f(x) - f(x_1)| < 1$, but there is no a priori reason to think that this δ is $\delta(1)$. The point we wish to make is that in the definition of continuity, given ϵ, the δ we choose is not only a function of ϵ but also a function of the point in question. So where we wrote $\delta(1)$ above, we would have more properly written $\delta(1,x_0)$; and for the second δ we could have written $\delta(1,x_1)$ where, in general, these are two different numbers.

We are led, therefore, to ask a question. Are there some functions for which, given any $\epsilon > 0$, a single value of δ will suffice, no matter what point in the domain of the function we consider? The answer is yes. Such functions are called uniformly continuous.

36.

Definition

A function f is uniformly continuous on a set S -- usually S will be some interval -- if, given any $\varepsilon > 0$, there exists a $\delta > 0$ such that $|f(x) - f(y)| < \varepsilon$ for all x and y in S such that $|x-y| < \delta$.

What we wish to emphasize here is that a single δ will do for the given ε; δ depends on ε but not on the points in S. We should also note that, although it is meaningful to speak of a function continuous at a point, uniform continuity at a point is a nonsense statement.

It follows almost immediately from the definition that the sum and the difference of uniformly continuous functions are uniformly continuous. Similarly, if we multiply a uniformly continuous function by a constant, the result is a uniformly continuous function.

However, the reader may be surprised to learn that the product of uniformly continuous functions need not be uniformly continuous. The standard example to illustrate this is the function $f(x) = x$. This function is clearly uniformly continuous on the entire real line (simply take $\delta = \varepsilon$). But $g(x) =$ $(f \cdot f)(x) = x^2$, although continuous on the real line, is not uniformly continuous. It is an easy exercise to show that the δ depends both on ε and on the point in question.

We now wish to discuss the central theorem on uniform continuity. However, we first state a theorem, important in its own right, which is used in the proof of this theorem.

The theorem is stated in terms of open and closed sets. On the real line, an open set is an open interval or the union of open intervals. You will note that every point in an open set S -- think of the interval (a,b) -- is contained in an open interval which itself is a subset of S. A set S is closed if its complement is open. Of course, a set need not be open or closed. Take, for example, the set [a,b). The entire real line and the null set are both open and closed, and these are the only such sets. Analogous definitions hold for higher dimensional spaces but we shall not discuss this point here.

Theorem 7. (Heine-Borel)

If a closed bounded set S is covered by a collection of open sets, then a finite number of open sets may be chosen from this collection in such a way that S is covered by this finite number of sets.

Every condition in the theorem is essential. For example, let S be (0,1). The collection of open sets (1/n, 1-1/n), n = 3,4,... , covers S, that is, every point in S is in at least one of these sets. But no finite number of this collection covers S. Here, of course, S is not closed. Again, the collection of open sets (-n,n), n = 1,2,..., covers S = R. But no finite number covers S since S is not bounded.

It is important that the collection of covering sets be open for the theorem to be true. Consider the closed bounded set S = [0,1]. Now every one point set of real numbers, {a}, is closed -- its complement is open. The collection of singletons consisting of the points in S forms a covering of S, where each set of this covering is closed. But there is no finite subcollection of this covering which will cover S.

The main theorem on uniform continuity is the following one.

Theorem 8.

Let f be defined and continuous at each point of a closed bounded set S. Then f is uniformly continuous on S.

That S must be closed is easily seen if we consider the function $f(x) = 1/x$. This function is defined and continuous on the bounded interval $(0,1]$, but it is not uniformly continuous on this interval. We might note, however, that this function is uniformly continuous on $(a,1]$, where $a > 0$. Hence the theorem describes a sufficient, not necessary, condition for uniform continuity.

Again, the set $S = R$ is closed and the function $f(x) = x^2$ is continuous on the entire real line. But this function is not uniformly continuous on S. Here, of course, S is not bounded.

As a final remark, we illustrated above that the product of two uniformly continuous functions need not be uniformly continuous. However, our example used uniformly continuous functions that were unbounded. If two functions f and g are both uniformly continuous and bounded, then their product can be shown to be uniformly continuous.

The reader may show that if a function f is uniformly continuous on a bounded interval, then the function is bounded. But the function $f(x) = x$ shows that the conclusion may be false if the interval fails to be bounded.

4. Some Theorems on Continuous Functions.

With the exception of uniform continuity, our concern to this point has been with local properties of continuous functions, -- continuity at a point, persistence of sign, etc. Local properties are generally more easily handled than global ones. However, there are important statements that can be made about continuous functions on closed intervals. These are the theorems we wish to discuss in this section. The proof of these theorems depends on the completeness of the real number system. To explain this, we start with some definitions.

Definition.

Let $S \subseteq R$ be a subset of real numbers. A number s is called an upper bound for S if, for all x in S, we have $x \le s$.

Definition.

A number **s** is called the least upper bound of S if

i. s is an upper bound of S;

ii. if s' is an upper bound of S, $s \le s'$.

The least upper bound of S will be denoted by sup S. (Some authors also use l.u.b. S.) The least upper bound of a set, sup S, should be distinguished from the maximum element of the set, max S. For example, the sets $(0,1)$ and $(0,1]$ have the same least upper bound, namely 1. But only the second set has a maximum element. The maximum element of a set always belongs to the set, whereas the least upper bound may or may not belong to the set as the above example indicates.

We will leave it to the reader to formulate the definition for the lower bound of a set and for the greatest lower bound, (denoted by inf S or g.l.b. S)

Not every set has an upper bound. For example, the positive integers or the set of all real numbers has no upper bound. On the other hand, the null set, $\emptyset$, has every real number as an upper (and lower) bound.

We are now in a position to formulate the completeness of the real number system. We do this in the following theorem.

Theorem 9.

Let $S \subset R$ be a nonempty set of real numbers which has an upper bound. Then S has a least upper bound i

R.

There are equivalent ways of stating the completeness of the real number system as we shall indicate in chapter 6. However, the above formulation is the one that is usually employed in proving the theorems that follow on continuous functions.

The reader should be aware that not every system of numbers is complete. Take, for example, the rational numbers and consider the nonempty set of rationals $S = \{x: x \in Q, x < 2^{\frac{1}{2}}\}$. This set has an upper bound. In fact it has infinitely many upper bounds. But there is no rational number that is a least upper bound of S. Thus, S has no least upper bound in the rationals. The rational numbers are not complete.

Theorem 10. The Intermediate Value Theorem.

Let f be continuous of a closed interval $[a,b]$, and suppose $f(a) < c < f(b)$. Then there is a point x in $[a,b]$ such that $f(x) = c$.

It is essential that f be continuous on the entire interval. Discontinuity at a single point can make the conclusion of the theorem false. For example, let $f(x) = \begin{cases} 1 & 0 \le x \le 1 \\ -1 & -1 \le x < 0 \end{cases}$. Here $f(-1) < 0 < f(1)$,

but there is no point in [-1,1] at which f(x) = 0.
The reason is that f is discontinuous at the origin.

The closed interval is an example of what is
called a connected set. The theorem only holds for
such sets. For instance, if g is a function continuous on the disjoint closed sets [a,b] and [c,d], the
intermediate value theorem is applicable to each set
separately, but not the union [a,b] $\cup$ [c,d].

Essentially the intermediate value theorem
states that if f is continuous on a closed interval
[a,b] and if f(a) $\neq$ f(b), then as x varies from a to
b, f(x) takes on every value between f(a) and f(b).
The reader should note that the theorem gives a
sufficient condition for f to take on every value
between f(a) and f(b). The condition is not necessary.
To see this, consider the function defined on [0,1]
by g(x) = $\begin{cases} x & x \text{ rational} \\ 1-x & x \text{ irrational} \end{cases}$.   This function is
discontinuous at every point in [0,1] except at x = ½
where it is continuous. Nevertheless, as x varies
from 0 to 1, g takes on every value between g(0) and
g(1).

Is there a converse to the intermediate value
theorem? By this we mean that if for any two points

$a < b$ and for any number c between $f(a)$ and $f(b)$ there is an x in $[a,b]$ such that $f(x) = c$, can we conclude that f is continuous. Plausible as this may sound, the conclusion is, in general, not true. In order to see this we need merely consider the

function $f(x) = \begin{cases} \sin(1/x) & -1 \leq x \leq 1, \ x \neq 0 \\ 0 & x = 0 \end{cases}$

The intermediate value theorem has many important applications. The reader might try to prove that any polynomial of odd degree has at least one real root by using this theorem.

Theorem 11.

If a function f is continuous on a finite closed interval $[a,b]$, it is bounded on that interval, that is, there exists a number M such that $|f(x)| \leq M$ for all x in $[a,b]$.

That the interval must be finite can be seen immediately by considering the function $f(x) = x$ on $[0,\infty)$. That the interval must be closed, we need only consider the function $g(x) = 1/x$ on $(0,1]$. Of course, discontinuous functions on closed bounded intervals can also be bounded. But the point is that they need not be so.

We end with a final theorem which gives a result somewhat stronger than that given by the previous theorem.

Theorem 12.

Let f be continuous on [a,b] and let m and M be the greatest lower bound and the least upper bound of the values of f(x) on [a,b]. Then f assumes each of the values m and M at least once in the interval.

Theorem 11 showed that the interval must be bounded. The continuity of the function is also essential. A function which is bounded on a closed finite interval certainly has a least upper bound and a greatest lower bound. But it does not follow that the function takes on these values in the interval. Consider the function

$$f(x) = \begin{cases} 1/1+x \quad \sin(1/x) & 0 < x \le 1 \\ 0 & x = 0 \end{cases}.$$

It is not difficult to see that $-1 < f(x) < 1$ on this interval and, in fact, sup $f(x) = 1$ and inf $f(x) = -1$ for $0 \le x \le 1$. However, there is no point x in the interval where the function actually takes on its maximum or minimum value.

If, rather than the above function, we consider $g(x) = 1/1+x \sin(1/x)$ where $0 < x \le 1$, we see that the requirement that the interval be closed cannot be omitted. This function is continuous and bounded on (0,1] but it does not attain its extreme values.

We end this section with a remark. In each of the above theorems we considered functions continuous on a closed bounded interval. As we have seen, such functions are uniformly continuous. But the wording, "continuous on a closed interval [a,b]" cannot be replaced by the wording, "a uniformly continuous function" in these theorems. We leave it to the reader to show this for each of the theorems by supplying examples.

III. DIFFERENTIATION.

1.   The Derivative.

Let f be a function defined on some set S. We
wish to study the difference quotient

$$1/h \{f(x_o+h) - f(x_o)\}$$

where, of course, we assume x_o+h and x_o are points
in S. In particular, we wish to investigate the
limit of this quotient as h approaches 0. In order
that the limit exist, the expression must make sense
Hence, we are led to make a preliminary definition.

Definition.

A point ξ is called a cluster point (or an accumulation point) of a set S if every open set containing ξ contains a point of S other than ξ.

First, it is to be noted that a cluster point
may, but need not, belong to the set itself. For example the set S = {1, 1/2, 1/3, 1/4,....} has 0 as a
cluster point but 0 does not belong to S. It follows
almost immediately from the definition that, if ξ
is a cluster point of S, then every open set containing ξ contains an infinite number of members of
S.

With these preliminary remarks made, we can pro-

48.

ceed to define the derivative of a function f at the point x_o.

Definition.

Let the domain of a function f be the set S and let the point x_o in S be a cluster point of S. Define the difference quotient of f at x_o by $1/h\{f(x_o+h) - f(x_o)\}$. If this quotient has a limit as h approaches 0, we say the function f is differentiable at x_o. This limit, which is denoted by $f'(x_o)$, that is,

$$f'(x_0) = \lim_{h \to 0} 1/h \{f(x_o+h) - f(x_o)\} ,$$

is called the derivative of f at x_o.

The domain of definition of f is generally an open set, a closed interval, etc. For such sets, every point is a cluster point. Of course, if the domain of definition of f is a closed interval or a half open half closed interval, we must understand the limit to be the one-sided limit. However, the definition we have given does not require that S be an interval. Thus, our definition also allows us to consider more general situations.

Now as we have remarked, a cluster point of a set need not belong to the set. However, in the definition of the derivative, we required that the cluster point belong to the set. The reason for this is clear. If we had not done so, the difference quotient would be undefined.

In order to see that this definition has meaning even if S is not an interval, consider a function f whose domain of definition is the set S = {1/n: n integer} $\cup$ {0}. Let f(x) = αx. The set S has only one cluster point, namely the origin. A straightforward application of the definition shows that f is indeed differentiable at the origin with f'(0) = α, that is

$$\lim_{\substack{h \to 0 \\ h \in S}} 1/h \{ f(h) - f(0)\} = \alpha.$$

This function is differentiable at no other point.

Now that we have made this point, however, unless it is otherwise stated explicitly, we shall consider all functions to be defined on some interval, so that every point of the domain of the function is a cluster point.

The first theorem on derivatives follows directly from the definition.

Theorem 1.

If f is differentiable at a point on its domain

x_o, then f is continuous at x_o.

Hence, if we consider the set of all continuous functions, the differentiable functions are a proper subset, since continuity is a necessary but not a sufficient condition for differentiability. The classic example to illustrate this is the absolute value function. Consider the function $f(x) = |x|$. This function is defined and continuous on the entire real line. However, it is not differentiable at the origin (it is at other points). If you look at the difference quotient, the limit is plus or minus one according as you approach the origin from the right or the left. Hence, the limit does not exist. Remember this example and you will never confuse differentiability and continuity in the theorem.

The reader may be thinking that this example is hardly that impressive. After all, the function does have a derivative everywhere except at one point. However, there are functions that are continuous at every point and differentiable at no point. Such functions are, as one would suspect, hardly like the rather simple functions we have encountered so far. The suspicion is correct. Nevertheless, we shall be able to give an example of an everywhere continuous and nowhere differentiable function but this must wait until we take up the consideration of series of functions in chapter 7.

We should emphasize that the limit of the difference quotient must be a finite number. For if we permitted infinite derivatives, then continuity would not be a necessary condition for differentiability. The function

$$f(x) = \begin{cases} x^{1/3} & x \leq 0 \\ a+x^{1/3} & x > 0 \end{cases},$$

where $a \neq 0$, has an "infinite derivative" at the origin but it is not continuous there.

Just as there are functions which fail to be differentiable at a single point, there are functions defined on an interval that are differentiable at one point only. We have previously considered the function

$$f(x) = \begin{cases} x & x \text{ rational} \\ 0 & x \text{ irrational} \end{cases}.$$

This function is continuous only at the origin, but it fails to be differentiable there or at any other point. However, consider the function

$$g(x) = \begin{cases} x^2 & x \text{ rational} \\ 0 & x \text{ irrational} \end{cases}$$

This function is again continuous only at the origin

52.

But it is also differentiable there and at no other point. The reader should have no difficulty in showing this by simply considering the definition of the derivative at a point.

If a function f has continuous derivatives on some interval, we can consider a new function, say g, whose value at $x_o + h$ is given by

$$g(x_o + h) = 1/h\{f'(x_o + h) - f'(x_o)\} \ .$$

If we now let h approach zero and if the limit exists we can define the function f" as

$$f''(x_o) = \lim_{h \to 0} g(x_o + h).$$

This is the second derivative of f evaluated at the point $x = x_o$. It is clear that we can define higher order derivatives in a similar fashion.

Of course, higher order derivatives need not exist. To see this, we need only consider a function which is differentiable everywhere but which has a discontinuous derivative. One such example would be

$$f(x) = \begin{cases} x^2 \sin(1/x) & x \neq 0 \\ 0 & x = 0 \end{cases} .$$

This function has a derivative everywhere but the derivative is discontinuous at the origin. Hence, f has no second derivative at the origin. Similarly, if we consider

$$g(x) = \begin{cases} x^4 \sin(1/x) & x \neq 0 \\ 0 & x = 0 \end{cases},$$

then g has a continuous derivative on the entire real line and the second derivative of g exists everywhere but is discontinuous at the origin. Hence, g fails to have a third derivative at the origin.

We should note, however, that the discontinuities in these examples are not simple discontinuities (discontinuities of the first kind). This is no coincidence. In fact, it can be shown that, if f is differentiable on an interval, its derivative, f', cannot have any simple discontinuities on the interval.

Theorem 2.

If f and g are differentiable at x_o, then f + g and f·g are differentiable at x_o.

The converse to this theorem, however, is easily seen to be false. We need only consider the functions

$$f(x) = \begin{cases} 1 & x \text{ rational} \\ -1 & x \text{ irrational} \end{cases} \quad \text{and} \quad g(x) = \begin{cases} 1 & x \text{ irrational} \\ -1 & x \text{ rational} \end{cases}$$

Neither function has a derivative at any point of the real line. But both the sum and the product of these

54.

functions, being constants, are differentiable everywhere.

Theorem 3.   (The Chain Rule)

Let g be differentiable at x_0 and let f be differentiable at $g(x_0)$. Then the composite function, $f \circ g$, is differentiable at x_0 and $(f \circ g)'(x_0) = f'(g(x_0)) \cdot g'(x_0)$.

In the theorem, both functions must be differentiable at the stipulated points, or else the conclusion of the theorem does not follow. In order to illustrate this, it is enough to consider the sine function and the absolute value function. Let g be the sine function and f the absolute value function. And let $x_0 = 0$. Then $f \circ g = |\sin|$. Here g is differentiable at $x = 0$ but f is not differentiable at $g(0) = 0$.

Again, let g be the absolute value function and f the sine function. Then $(f \circ g)(x) = \sin|x|$.   In this case f is differentiable at $g(0) = 0$, but g is not differentiable at the origin.

We must emphasize that f is to be differentiable at $g(x_0)$ not at x_0. For instance, in the example above, where $f \circ g = |\sin|$, if we take $x_0 = \pi$, then f and g are differentiable at that point but, since $g(\pi) = 0$ and f is not differentiable at the origin, the composite function is not differentiable at $x_o = \pi$.

Theorem 4.

Let f be differentiable at x_o and suppose $f(x_o) \neq 0$. Then $|f|$ is differentiable at x_o and

$$|f|'\ (x_o) = f'(x_o) \cdot |f(x_o)| / f(x_o).$$

It is clear, however, that the differentiability of $|f|$ at a point need not imply the differentiability of f at that point as either function used in the example after theorem 2 shows. Again, $f(x_o)$ must not vanish. Simply consider $f(x) = x$ at $x_o = 0$.

2. Maxima and Minima.

We have already seen that, if a function is continuous on a closed bounded interval, then the function takes on a greatest and a least value somewhere on that interval. We now wish to investigate the method of finding the points where the function takes on its maximum and its minimum values. Since we will be discussing both absolute extrema and local extrema on a set, we should begin by distinguishing these.

Definition.

Let f be a function defined on some set S. If

for some point x_o in S, $f(x_o) \geq f(x)$ for all x in S, we say that $f(x_o)$ is the absolute maximum value of f in S and that f takes on its absolute maximum value in S at $x = x_o$. Similarly, if for some x_o in S, $f(x_o) \leq f(x)$ for all x in S, we say that $f(x_o)$ is the absolute minimum value of f in S and that f takes on its absolute minimum value in S at $x = x_o$.

A function need not take on its absolute maximum value or its absolute minimum value at a single point. Consider the sine function on the entire real line. Again, the absolute maximum value and the absolute minimum value of the function may be the same as with the constant function on some interval.

We distinguish absolute extrema from relative or local extrema.

Definition.

Let f be defined on an open interval (a,b), and let x_o be a point of this interval. The function f has a relative (local) maximum at x_o if there is some open interval containing x_o, say (a_1,b_1), where $a \leq a_1 < x_o < b_1 \leq b$, such that $f(x) \leq f(x_o)$ for all x in (a_1,b_1). The function f has a relative (local) minimum at x_o if there is some open interval containing x_o, say (a_2,b_2), where $a \leq a_2 < x_o < b_2 \leq b$, such

57.

that $f(x_0) \leq f(x)$ for all x in (a_2, b_2).

The following theorem is the most important one on extreme values of a function.

Theorem 5.

Let f be defined on an open interval (a,b). If f takes on a maximum or a minimum value (absolute or relative) at a point x_0 in (a,b) and if f is differentiable at x_0, then $f'(x_0) = 0$.

Note that the theorem holds for absolute and relative extrema alike. Again, the function is to be differentiable at the point in question. A function can take on a maximum or a minimum value at a point where its derivative does not exist. The theorem says nothing about this.

The function defined on (-1,1) by

$$f(x) = \begin{cases} 0 & x \text{ rational} \\ 1 & x \text{ irrational} \end{cases}$$

takes on its minimum at the origin and at all rational points, but this function is not differentiable at any point. The function defined on (-1,1) by $g(x) = \begin{cases} x^2 & x \text{ rational} \\ 0 & x \text{ irrational} \end{cases}$

has a derivative only at the origin where $g'(0) = 0$. This function takes on its minimum value at the origin where the derivative exists and at all irrational points of the interval where the derivative does not exist. Note that in both these examples the extrema were absolute extrema.

The theorem gives a necessary condition for maximum or minimum, but the condition is not sufficient. Just consider the function $f(x) = x^3$ on $(-1,1)$. Its derivative vanishes at the origin but the origin is neither maximum nor minimum point. Again, the domain of the function is to be an open interval. On the closed interval $[0,1]$ the function $f(x) = x$ takes on a maximum at $x = 1$, but its derivative (one-sided) does not vanish at that point. Basically, what the theorem is saying is that if the derivative of a function exists and is unequal to zero at some point of an open interval, that point cannot be a maximum or a minimum point.

If we look upon the derivative as giving the slope of the tangent to the curve given by some continuous function, our geometric intuition should be enough to convince us of the truth of the following theorem.

Theorem 6.

Suppose f is continuous on some open interval (a,b) with x_o a point of this interval, and suppose for all points of (a,b) except possibly x_o, the derivative f' exists.

1. If $f'(x) > 0$ for all x in some open interval having x_0 as a right end point and if $f'(x) < 0$ for all x in some open interval having x_0 as a left endpoint, then f has a relative maximum value at x_0.

2. If $f'(x) < 0$ for all x in some open interval having x_0 as a right end point and if $f'(x) > 0$ for all x in some open interval having x_0 as a left endpoint, then f has a relative minimum value at x_0.

We emphasize that the theorem says nothing about the derivative of f at x_0. The derivative need not even exist at x_0. The most obvious example of this is $f(x) = |x|$ with $x_0 = 0$. The theorem shows that the origin is a minimum point of this function, in this case an absolute minimum, even though this function is not differentiable at the origin.

Using a superficial geometric reasoning, the reader may be tempted to think that the theorem can be turned around in the other direction. For instance we might argue that if f is continuous and differentiable in some interval (a,b) and if f takes on a relative minimum at some point x_0 in this interval, then there is some open set with x_0 as its right endpoint on which $f'(x) < 0$ in this set and some open

set with x_o as its left endpoint on which $f'(x) > 0$.
This is not so! For consider the function

$$f(x) = \begin{cases} x^2 \ (1+\sin^2(1/x)) & x \neq 0 \\ 0 & x = 0 \end{cases}.$$

For this function we have

$$f'(x) = \begin{cases} 2x(1+\sin^2(1/x)) - 2 \sin(1/x) \cdot \cos(1/x) & x \neq 0 \\ 0 & x = 0 \end{cases}.$$

It is clear from the nature of this function that
$x = 0$ is its absolute minimum. Yet there is no open interval, no matter how small, with the origin as left
endpoint on which f' is positive nor is there any open
interval with the origin as right endpoint on which f'
is negative. The reader might find it a worthwhile exercise to roughly plot the graph of this function in
order to convince himself of this fact.

The careful reader may have noticed that the derivative of this function is discontinuous at the origin. But this has nothing to do with the argument. If
we had used an initial factor of x^4 rather than x^2,
the derivative would then be continuous at the origin.
Yet the function would have essentially the same behavior.

Points at which the derivative of a function vanish are called critical points. Now suppose we have a
function defined on some interval. How do we find the

points at which the function takes on its maximum and minimum values? If the interval is open and if the function is differentiable on the entire interval, theorem 5 gives the answer. Investigate the critica points! But always keep in mind that not all critical points need be extreme points.

In the more likely situation in which the function may not be differentiable everywhere, or where the domain of the function may be a closed interval or a half open half closed interval, the maximum and minimum points may be found by a three step process. Say we wish to find the relative extreme points of f on [a,b]. Investigate:

1. Any critical points of f on (a,b).
2. Points in (a,b) where f is not differentiable.
3. The endpoints a and b.

In general, not all these points will be extreme points, but any extreme points that do exist will be found among these points.

3. The Mean Value Theorem.

One of the most important theorems in differential calculus is the mean value theorem. Most books first consider Rolle's theorem from which the mean value theorem easily follows.

Theorem 7. (Rolle's Theorem)

Let f be continuous on a closed interval [a,b] and differentiable on the open interval (a,b). Suppose f(a) = f(b). Then there exists a point x_o in (a,b) such that $f'(x_o) = 0$.

It is essential that the interval on which f is continuous be closed. For example, let f(x) = x on [0,1) and let f(1) = 0. Then f(0) = f(1), f is differentiable on (0,1), and f is continuous on [0,1). But there is no point x_o in (0,1) for which $f'(x_o) = 0$.

Theorem 8. (The Mean Value Theorem)

Let f be continuous on a closed interval [a,b] and differentiable on the open interval (a,b). Then there is a ξ in (a,b) such that

$$f(b) - f(a) = f'(\xi) \cdot (b-a).$$

The example after Rolle's theorem shows that f must be continuous on the entire closed interval. Again, f must be differentiable at every point of the open interval. For consider $f(x) = x^{2/3}$ on [-1,1]. This function is certainly continuous on this interval. But there is no ξ in (-1,1) such that f(1) - f(-1) = $f'(\xi) \cdot (1-(-1))$, that is, no point such that $f'(\xi) = 0$. The mean value theorem fails here, of course, because f is not differentiable at the origin (it is at other points in (-1,1)).

The mean value theorem can be generalized as follows.

Theorem 9. (Cauchy's Generalized Law of the Mean)

Let f and g be continuous on the closed interval [a,b], and differentiable on the open interval (a,b). Then there is a ξ in (a,b) such that

$$(f(b) - f(a)) \cdot g'(\xi) = (g(b) - g(a)) \cdot f'(\xi).$$

The reader should note that if g(x) = x, this theorem reduces to the mean value theorem. However, this generalization is truly new. For by applying the mean value theorem to f and g individually, we could derive the relation

$$(f(b) - f(a)) \cdot g'(\xi) = (g(b) - g(a)) \cdot f'(\zeta),$$

but we would have no justification for setting ξ = ζ.

For those with a knowledge of determinants, this theorem can be further generalized if we consider func-tions f, g, and h which satisfy the conditions of theorem 9. In this case, there exists a ξ in (a,b) such that

$$\begin{vmatrix} f(a) & g(a) & h(a) \\ f(b) & g(b) & h(b) \\ f'(\xi) & g'(\xi) & h'(\xi) \end{vmatrix} = 0$$

The reader should be able to show that theorems 8 and 9 are special examples of this result.

4. Monotonic Functions.

Definition.

Let f be defined on some interval. The nature of the interval need not be specified.

1. f is said to be increasing on the interval if $f(y) \leq f(x)$ for all x and y in the interval such that $y < x$. If strict inequality holds for all such x and y, the function is said to be strictly increasing on the interval.

2. f is said to be decreasing on the interval if $f(y) \geq f(x)$ for all x and y in the interval such that $y < x$. If strict inequality holds for all such x and y, the function is said to be strictly decreasing on the interval.

Definition.

A function which is either (strictly) increasing or (strictly) decreasing on an interval is said to be (strictly) monotonic on the interval.

We must distinguish functions increasing or decreasing on an interval from functions increasing or decreasing at a point.

Definition.

A function is increasing at a point x_o if there is some open interval with x_o as its right end point.

say (a,x_0), such that $f(x) < f(x_0)$ for all x in (a,x_0), and some open interval with x_0 as its left endpoint, say (x_0,b), such that $f(x_0) < f(x)$ for all x in (x_0,b).

A similar definition holds for a function decreasing at a point. As the following example illustrates, a function increasing at the point x_0 is not necessarily increasing on an open interval containing x_0. Let

$$f(x) = \begin{cases} x + x^{4/3} \sin(1/x) & x \neq 0 \\ 0 & x = 0 \end{cases}.$$

It is not difficult to see that this function is increasing at the origin. However, on any open interval containing the origin, the function is not monotonic. Its derivatives take on both positive and negative values.

But we do have the theorem.

Theorem 10.

If $f'(x_0) > 0$ and f' is continuous at x_0, then f is increasing on some interval containing x_0.

But the previous example illustrates that it is not enough that $f'(x_0) > 0$. In this example $f'(0) = 1$, but the function is not increasing. Why? Because the derivative is not continuous at the origin.

What can be shown is the following.

66.

Theorem 11.

If f is continuous and differentiable on an interval and has no critical points on the interval except, perhaps, the endpoints, then f is strictly monotonic on the interval.

The reader should have no difficulty in supplying an example to show that if the differentiability of f is omitted, then the function need not be monotonic.

5. Sufficient Conditions for Maximum and Minimum.

We have seen that if a function is defined and differentiable on some open interval and if the function takes on an extreme value at some point x_0 of the interval, it is necessary that $f'(x_0)= 0$. But this condition hardly guarantees that the point x_0 will be an extreme point. We cited the example where $f(x)= x^3$. Then $f'(0)= 0$, yet the origin is not an extreme point. We are, therefore, led to inquire into sufficient conditions for relative extrema.

Theorem 12.

Let $f'(x_0)= 0$ and let f' be defined for all x in some interval containing x_0.

1. If $f''(x_0) < 0$, $f(x_0)$ is a relative maximum of f.

2. If $f''(x_0) > 0$, $f(x_0)$ is a relative minimum of f.

3. If $f''(x_0) = 0$, no conclusion can be drawn.

In order to convince yourself that no conclusion can be drawn if $f''(x_0) = 0$, consider the functions $f(x) = x^3$ and $g(x) = x^4$. The origin is a critical point for both functions and $f''(0) = g''(0) = 0$. But whereas the origin is a minimum point for g, it is neither a maximum nor a minimum point for f.

We can see from the theorem that if f has a relative minimum at x_0 and the second derivative exists at this point, we can conclude that $f''(x_0) \geq 0$. However, because of part three of the theorem, we cannot conclude strict inequality. Similar remarks hold for a relative maximum at x_0.

Now suppose the second derivative vanishes? Must we despair of finding a sufficient condition for relative extrema? The answer is no. Although, often enough in such an event, geometric considerations may be of more use than higher derivative tests.

Theorem 13.

Suppose f and its first n derivatives are defined for all x in some interval containing x_0, and suppose $f'(x_0) = f''(x_0) = \ldots = f^{(n-1)}(x_0) = 0$. But suppose $f^{(n)}(x_0) \neq 0$. Assume, moreover, that $f^{(n)}$ is continuous at x_0. Then for n even:

1. If $f^{(n)}(x_0) < 0$, $f(x_0)$ is a relative maximum value of f.

2. If $f^{(n)}(x_o) > 0$, $f(x_o)$ is a relative minimum value of f.

3. If n is odd, there is neither relative maximum nor relative minimum of f at x_o.

6. Convex Functions.

We have seen that the sign of the derivative of a function in some interval gives some indication of the behavior of the function -- increasing, decreasing, oscillating, etc. Does the sign of the second derivative lend itself to a similar geometric interpretation? We shall see that the answer is affirmative.

Consider a function f, continuous with continuous first and second derivatives, for which the second derivative is everywhere positive on some interval. Then the first derivative must be strictly increasing on this interval. And so the slope of the tangent to the curve is increasing from left to right. Geometrically, this means that the curve is bending upwards. The parabola given by $f(x) = x^2$ is the most simple illustration of this. If, on the other hand, the second derivative is negative on some interval, the function f must represent a curve bending downwards on the interval as with $f(x) = -x^2$.

Points at which the second derivative of a function change sign from positive to negative or from negative to positive are called points of inflection. At such points the second derivative of the function vanishes. However, it is not to be thought that every point at which the second derivative vanishes is an inflection point. For example, let $f(x) = x^4$. The second derivative vanishes at the origin, but is positive both to the left and to the right of the origin. Hence, the origin is not an inflection point. But we do have the following result.

Theorem 14.

If $f''(x_o) = 0$ and $f'''(x_o) \neq 0$, then the graph of the function f has a point of inflection at x_o.

An example of a function which has such behavior would be $f(x) = x^3$ at $x = 0$. But again, $f'''(x_o) \neq 0$ is only a sufficient condition that x_o be an inflection point. For the function $f(x) = x^5$, the third derivative vanishes at the origin along with the second derivative. Yet for this function, the origin is still a point of inflection.

Definition.

A function f is convex on some interval I if

$$f((1-\lambda)x + \lambda y) \leq (1-\lambda)f(x) + \lambda f(y)$$

for all x and y in the interval and for $0 \leq \lambda \leq 1$.

The reader should not confuse a convex function

with a convex set, that is, a set which contains the line segment joining any two of its points. However, the two concepts are not unrelated.

Theorem 15.

A necessary and sufficient condition that a function f be convex on an interval I is that the set of points above the graph of the function, $S = \{(x,y): x \in I, y \geq f(x)\}$, be a convex set.

There is another definition of convexity that is often useful.

Definition.

A function f is convex on an interval I if for any points x_0, x_1, x_2 in the interval with $x_1 < x_0 < x_2$ we have

$$\frac{f(x_0) - f(x_1)}{x_0 - x_1} \leq \frac{f(x_2) - f(x_1)}{x_2 - x_1} \quad .$$

We should explore more carefully the geometric implications of convexity. We shall do so in the following theorems.

Theorem 16.

Let f be a convex function which is differentiable on an open interval. Then the graph of f lies above the tangent line to f at all points of the interval. Moreover, if f is differentiable at x_1 and x_2, with $x_1 < x_2$, then $f'(x_1) \leq f'(x_2)$.

71.

It is essential that we specify the differentiability of f at the points in question since it is no
necessary that a convex function be differentiable a
every point. For example, the function f(x) = |x| i
convex but is not differentiable at the origin. Ther
is a converse to theorem 16.

Theorem 17.

If f is differentiable and the graph of f lies
above each tangent line to f except at the point of
contact with the tangent line, then f is convex.
Moreover, if f is differentiable and if f' is increasing, then f is convex.

All the statements made about convex functions
have analogies for what are called concave functions
We leave it to the reader to formulate the appropriate definitions and theorems. We only remark that,
f is convex on some interval I, then -f is concave o
the same interval.

We will end with a brief discussion of the connection between convexity and the second derivative
of the function f. It follows from theorem 16 that,
if f is convex on an open interval I, then f" is non
negative at all points of the interval where the
second derivative exists. Moreover, if f" exists an

is nonnegative on an open interval I and if f" is not identically zero on any subinterval of I, then f is convex on the interval.

With these results and similar ones for concave functions, we can now analyze the geometric nature of an inflection point of the graph of a function f. If a function is convex on an interval, the tangent line to the graph of the function always lies below the graph of the function. Similarly, if f is concave on an interval, the tangent line to the graph of the function always lies above the graph of the function. A point of inflection of the graph of a function is precisely that point at which the tangent line to the graph of the function crosses the graph. The most simple illustration of this **is** $f(x) = x^3$ at the origin.

We end our discussion with a final theorem on convex functions.

Theorem 18.

If f is convex (concave) on an open interval I, then f is continuous on I.

We note, however, that if "open interval" is replaced by "closed interval", the conclusion of the theorem need not hold. For instance, consider the function

$$f(x) = \begin{cases} 1 & x = 0 \\ x^2 & x > 0 \end{cases} .$$

This function is convex but it is not continuous at the origin.

It should not have to be remarked that continuous functions need not be convex or concave.

IV.   THE INTEGRAL.

1.   Definition of the Integral and Its Properties.

In the study of integration, the novice in
calculus must cope with two problems. First, the
integral must be defined and its various properties
explored. Then there is the question of the techniques of integration. For a given function, what
is the integral of the function? Unfortunately, it
is often the case that the student becomes too pre-
occupied with techniques in a first course, with the
result that the definition of the integral is never
really understood.

We will not be concerned with the techniques
of integration in this chapter. They are standard
and can be found in any calculus text book. Rather
we shall concentrate on the definition of the integral and its principal properties. The approach
will be analytic rather than geometric. The reason
for this is not to belittle geometric insight. To
look upon the integral as the area under a curve
can often be helpful. But, as we shall see, the
class of integrable functions includes many functions
for which we would be hard-pressed to assign any
geometric picture.

Definition.

Consider the closed bounded interval [a,b]. A partition, P, of [a,b] is a finite set of points in the interval, $\{x_0, x_1, \ldots, x_n\}$, such that

$$a = x_0 < x_1 < \ldots < x_{n-1} < x_n = b.$$

Now take a bounded real valued function f, defined on the interval [a,b] and make a partition, P, of [a,b]. For convenience, we will use the notation $\Delta x_k = x_k - x_{k-1}$ (k = 1,2,...,n). We write

$$M_k = \sup f(x) \qquad x \in \Delta x_k$$

$$m_k = \inf f(x) \qquad x \in \Delta x_k$$

and form what we call the upper and lower sums of f with respect to the partition P,

$$U(P,f) = \sum_{k=1}^{n} M_k \Delta x_k \quad ,$$

$$L(P,f) = \sum_{k=1}^{n} m_k \Delta x_k \quad .$$

Note that the upper and lower sums depend both on the function f and the partition P. We use sup and inf rather than max and min since there is no requirement that the function be continuous. The function need not attain its maximum or minimum value on the interval. Since the function and the interval are both bounded, the numbers U(P,f) and L(P,f) are bounded as we vary the partition. Hence, we can take the infirmum and the supremum of these numbers. By doing so, we obtain the upper and lower Riemann integrals.

Definition.

By the upper Riemann integral is meant the number

$$\inf U(P,f) = \int_a^{-b} f\ dx,$$

and by the lower Riemann integral is meant the number

$$\sup L(P,f) = \int_{-a}^{b} f\ dx$$

where the inf and sup are taken over all partitions of [a,b].

The upper and lower integrals quite definitely exist. However, they need not be equal. For example. on the interval [0,1] let

$$f(x) = \begin{cases} 1 & x\ \text{rational} \\ 0 & x\ \text{irrational} \end{cases} .$$

It is not difficult to see that

$$\int_0^{-1} f \, dx = 1 \quad \text{and} \quad \int_{-0}^{1} f \, dx = 0.$$

In order to see how the upper and lower sums depend on the partitions, we make a definition.

Definition.

Let P and P' be two partitions of an interval [a,b]. If every point of P is a point of P' and if P ⊂ P' (proper inclusion), the partition P' is said to be a refinement of the partition P.

If we start with a partition P and make a refinement P' of P, for example by adding a single point to P, a little thought shows that the lower sums increase and the upper sums decrease, that is $L(P',f) \geq L(P,f)$ and $U(P',f) \leq U(P,f)$.

We can say more. It is easy to see that given a partition P, we have $L(P,f) \leq U(P,f)$. Not only that but any upper sum is an upper bound for every lower sum, and any lower sum is a lower bound for every upper sum. That is to say, for arbitrary partitions P_1 and P_2, we have

$$L(P_1,f) \leq U(P_2,f).$$

To show this, we need only consider the partition

$P = P_1 \cup P_2$. This means that P contains all the points belonging to P_1 and all the points belonging to P_2. We then have $L(P_1,f) \leq L(P,f) \leq U(P,f) \leq U(P_2,f)$. From this it follows that the upper Riemann integral is equal to or greater than the lower Riemann integral,

$$\int_{-a}^{b} f \, dx \leq \int_{a}^{-b} f \, dx.$$

We are now in a position to define the Riemann integral.

Definition.

A real valued bounded function f, defined on an interval [a,b], is Riemann integrable on [a,b] if and only if

$$\int_{-a}^{b} f \, dx = \int_{a}^{-b} f \, dx.$$

The common value of the upper and lower Riemann integral is called the Riemann integral and is denoted by

$$\int_{a}^{b} f \, dx.$$

That not every bounded function is Riemann integrable is obvious from the example given above. The function in that example is, of course, discontinuous at every point in the interval. But it is not to be thought that integrability requires continuity. We shall show below that even some badly discontinuous functions have equal upper and lower Riemann integrals and, hence, are integrable.

We might mention here that there is a more general definition of integral due to Lebesgue. Even the everywhere discontinuous function in the example above is integrable in the sense of Lebesgue. It can be shown that every properly Riemann integrable function is Lebesgue integrable, and both integrals have the same value. But the Lebesgue integral is beyond the scope of this work.

Our first theorem involves little more than putting the definition of integrability in a different form.

Theorem 1.

A bounded real valued function f is integrable on the interval [a,b] if and only if for every $\varepsilon > 0$, there exists a partition P of the interval such that

$$U(P,f) - L(P,f) < \varepsilon.$$

This theorem can be used to show that continuity is not necessary in order that a function be integrable. For example, consider the bounded function f, defined on [-1,1] by $f(x) = \begin{cases} 0 & x \neq 0 \\ 1 & x = 0 \end{cases}$.

If we take any partition and merely require that the subinterval containing the origin, say Δx_k, be of length less than ε, then the conditions of the theorem are satisfied, f is integrable and

$$\int_{-1}^{1} f \, dx = 0.$$

However, continuity of a function on an interval [a,b] is a sufficient condition for the Riemann integrability of the function. Recall that a function f, continuous on [a,b], is automatically bounded on that interval.

Theorem 2.

If f is continuous on the interval [a,b], then f is Riemann integrable on the interval.

Thus, we have a whole class of integrable functions. But as the example shows above, the continuous functions do not exhaust the class of integrable functions.

The monotonic functions defined on an interval [a,b] furnish a second class of integrable functions This is easily seen to be so by making a uniform partition of the interval and looking at the difference between the upper and lower sums. An immediate application of theorem 1 leads to the desired conclusion. We state this result as a theorem.

Theorem 3.

If f is monotonic on the interval [a,b], then f is Riemann integrable on the interval.

It should hardly be necessary to remark that the class of continuous functions on some interval and the class of monotonic functions on the same interval have some members in common but, in general, are quite distinct.

At the risk of confusing the reader, let us briefly introduce a third class of Riemann integrabl functions, a class not ordinarily treated in a first course in calculus.

Definition.

Let f be a real valued function defined on an interval [a,b], and let $P = \{x_0, x_1, \ldots, x_n\}$ be a partition of [a,b]. Let $\Delta f_k = f(x_k) - f(x_{k-1})$.

Define the total variation of f on [a,b] to be

$$V(f;[a,b]) = \sup \sum_{k=1}^{n} |\Delta f_k|$$

where the supremum is taken over all partitions of
[a,b]. The function f is of bounded variation on
[a,b] if and only if $V(f;[a,b]) < \infty$.

Theorem 4.

If f is of bounded variation on the interval
[a,b], then f is Riemann integrable on the interval.

It is not difficult to see that every monotonic
function is of bounded variation. This is not true,
however, of continuous functions. The function

$$f(x) = \begin{cases} x \sin(1/x) & 0 < x \leq 1 \\ 0 & x = 0 \end{cases}$$

is continuous on [0,1] and, hence, is integrable
over the interval. But this function is not of
bounded variation. (Most readers will not be able
to prove the latter statement at this stage, but
after chapter 6, they should be able to show this.)

We now list some elementary theorems on Riemann
integrable functions with appropriate examples where

they are required.

Theorem 5.

If f_1 and f_2 are Riemann integrable on $[a,b]$, then $f_1 + f_2$ is Riemann integrable on $[a,b]$, and

$$\int_a^b (f_1+f_2)\ dx = \int_a^b f_1\ dx + \int_a^b f_2\ dx.$$

An example to show that the Riemann integrability of $(f_1 + f_2)$ does not imply that both f_1 and f_2 are Riemann integrable should quickly come to mind. Simply let f_1 and f_2 be defined on $[0,1]$ by

$$f_1(x) = \begin{cases} 1 & x \text{ rational} \\ -1 & x \text{ irrational} \end{cases} \quad \text{and}$$

$$f_2(x) = \begin{cases} 1 & x \text{ irrational} \\ -1 & x \text{ rational} \end{cases}.$$

If, however, $(f_1 + f_2)$ and, say, f_1 are Riemann integrable on an interval, the following theorem shows that f_2 is Riemann integrable on the interval.

Theorem 6.

If f is Riemann integrable on $[a,b]$, and if c is an arbitrary constant, then cf is Riemann integrable on $[a,b]$ and $\int_a^b cf\ dx = c \int_a^b f\ dx$.

84.

If we suppose that cf is Riemann integrable on an interval, is f Riemann integrable on the same interval? The reader should not be too quick to respond. The answer is yes -- as long as $c \neq 0$! Just consider the functions after theorem 5 to see that a false conclusion can be reached if we let $c = 0$.

Theorem 7.

If f is Riemann integrable on [a,b], and if $a < c < b$, then f is Riemann integrable on [a,c] and on [c,b], and

$$\int_a^b f \, dx = \int_a^c f \, dx + \int_c^b f \, dx.$$

Theorem 8.

If f is Riemann integrable on [a,b], then

$$\int_a^b f \, dx = - \int_b^a f \, dx, \quad \text{and} \quad \int_a^a f \, dx = 0.$$

Theorem 9.

If f_1 and f_2 are Riemann integrable on [a,b], then $f_1 \cdot f_2$ is Riemann integrable on [a,b].

The examples after theorem 5, however, show that $f_1 \cdot f_2$ can be Riemann integrable on an interval with neither f_1 nor f_2 Riemann integrable on the interval.

Moreover, even if $f_1 \cdot f_2$ and, say, f_1 are Riemann

integrable, f_2 need not be. In order to see this,

take $f_1 = 0$ and f_2 as above. Or, if you wish to

avoid a function vanishing on an interval, consider

the functions $f_1(x) = x$ and $f_2(x) = 1/x$ on $[-1,1]$.

Then $f_1 \cdot f_2 = 1$ and is integrable on the interval, as

is f_1, But f_2, being unbounded, has no proper Riemann

integral.

Theorem 10.

   If f_1 and f_2 are Riemann integrable on $[a,b]$,

and if $f_1(x) \leq f_2(x)$ for all x in $[a,b]$, then

$$\int_a^b f_1 \; dx \leq \int_a^b f_2 \; dx.$$

   If you think of some of the examples already

seen in this chapter, it should not be difficult to

convince yourself that the equality of the integrals

need not imply that the functions which are being

integrated are equal. In fact, the function we shall

consider after theorem 12 will show that the func-

tions may differ from each other at an infinite num-

ber of points, yet the integrals may be equal. And

again, from the inequality between the integrals, we cannot conclude that a corresponding inequality holds between the functions. We leave it to the reader to furnish examples.

Theorem 10 leads us to consider another inequality with integrals. Suppose that f is integrable on an interval I_1 and that the interval I_2 is a proper sub-interval of I_1, $I_2 \subset I_1$. Does it follow that

$$\int_{I_2} f \, dx \leq \int_{I_1} f \, dx \, ?$$

It does -- provided we put one further condition on f. The function f must be nonnegative. Otherwise, the inequality may not hold. Just consider the function $f(x) = x$ with $I_1 = [-1,1]$ and $I_2 = [0,1]$ in order to see this.

Theorem 11.

If f is Riemann integrable on [a,b], then |f| is Riemann integrable on [a,b], and

$$\left| \int_a^b f \, dx \right| \leq \int_a^b |f| \, dx.$$

Strict inequality can occur. Consider $f(x) = x$ on [-1,1]. The functions considered after theorem 5 show that the integrability of |f| does not imply the integrability of f.

The question now arises about the Riemann integrability of the composition of two Riemann integrable functions. Is the composite function Riemann integrable? In general, the answer is no!

Theorem 12.

Let f be Riemann integrable on [a,b] with M and m the upper and lower bound respectively of f on [a,b]. Let g be continuous on [m,M]. Then the composite function h = g∘f is Riemann integrable on [a,b].

Without the continuity of g, the composite function need not be integrable. We first exhibit a function considered earlier, a function that is continuous at every irrational point and discontinuous at every rational point. Let the function f be defined on [0,1] by

$$f(x) = \begin{cases} 0 & x \text{ irrational} \\ 1/q & x = p/q \text{ in lowest terms} \end{cases}$$

Even though discontinuous at infinitely many points, the rationals in the interval, this function is Riemann integrable in the interval, $\int_0^1 f \, dx = 0$.

This statement will be justified after theorem 17.
Now define a function g on [0,1] by

$$g(x) = \begin{cases} 0 & x = 0 \\ 1 & 0 < x \leq 1 \end{cases}.$$ Note that g is continuous

everywhere except at the origin so that g is certainly Riemann integrable. But the composite function is
not Riemann integrable. In this case we get

$$h(x) = (g \circ f)(x) = \begin{cases} 0 & x \text{ irrational} \\ 1 & x \text{ rational} \end{cases}.$$

So we see that the composition of two Riemann
integrable functions need not be Riemann integrable.
We could go further and give an example to show that
we cannot reverse the order of composition in the
theorem. That is to say, if g is Riemann integrable
and if f is continuous, the composite function $g \circ f$
need not be Riemann integrable. But to construct
such an example would take us beyond the level we
wish to set for this book. (We refer the reader to
Gelbaum and Olmsted's "Counterexamples in Analysis,"
p. 106.)

We close this section with two mean value
theorems for integrals.

Theorem 13.

Let f be Riemann integrable on [a,b], and let

f be bounded above by M and bounded below by m on the interval. Then there exists some number μ, $m \leq \mu \leq M$ such that $\int_a^b f\, dx = \mu(b - a)$. If, moreover, f is continuous on $[a,b]$, there exists a point x_0 in $[a,b]$ such that $\int_a^b f\, dx = f(x_0)(b - a)$.

In the second part of the theorem, it is essential that f be continuous on the interval. For consider the function $f(x) = \begin{cases} 0 & 0 \leq x < \frac{1}{2} \\ 1 & \frac{1}{2} \leq x \leq 1 \end{cases}$.

Here $\int_0^1 f\, dx = \frac{1}{2}$ but there is no x_0 in $[0,1]$ such that $f(x_0) \cdot (1-0) = \frac{1}{2}$.

Theorem 14.

Let f be continuous on $[a,b]$ and let g be Riemann integrable and nonnegative on $[a,b]$. Then there is a point x_0 in $[a,b]$ such that

$$\int_a^b f \cdot g\, dx = f(x_0) \cdot \int_a^b g\, dx.$$

Here the nonnegative character of g is to be emphasized. For take $f(x) = g(x) = x$ on $[-1,1]$. Then $\int_{-1}^1 f \cdot g\, dx = 2/3$ and $\int_{-1}^1 g\, dx = 0$. Clearly there is no x_o in $[-1,1]$ such that $f(x_o) \cdot 0 = 2/3$.

2.   The Fundamental Theorem of Calculus.

What is called "the fundamental theorem of calculus" is stated in a variety of different forms. Perhaps the most common is the following one.

Theorem 15.

If f is Riemann integrable on the interval $[a,b]$ and if there is a differentiable function F defined on $[a,b]$ such that $F' = f$, then

$$\int_a^b f\ dx = F(b) - F(a).$$

The function F with the properties stated in the theorem is called the primitive of f. It should be obvious that the primitive of a function is not unique. However, if F and G are both primitives of f on an interval, they are not unrelated. In this event, they can differ from each other only by a constant.

It is not to be thought that every integrable function has a primitive. For example, the function f defined on $[0,1]$ by $f(x) = \begin{cases} 0 & x \neq \frac{1}{2} \\ 1 & x = \frac{1}{2} \end{cases}$ is certainly integrable, but there is no differentiable function F such that F' = f on the interval. Or an even more dramatic illustration of this fact can be seen in the example after theorem 12.

Moreover, a function may possess a primitive without being Riemann integrable. For consider the function f defined on [0,1] by

$$f(x) = \begin{cases} 3/2 \cdot x^{\frac{1}{2}} \cdot \sin(1/x) - x^{-\frac{1}{2}} \cdot \sin(1/x) & x \neq 0 \\ 0 & x = 0 \end{cases}.$$

This function does, indeed, have a primitive on [0,1] namely

$$F(x) = \begin{cases} x^{3/2} \sin(1/x) & x \neq 0 \\ 0 & x = 0 \end{cases}, \text{ and } F' = f$$

on the interval. Yet f is not integrable since it is unbounded on the interval.

Up to this point we have considered only what is called the definite integral. We now turn our attention to the indefinite integral, or an integral with a variable upper (or lower) limit. Suppose f is Riemann integrable on an interval [a,b] and let x be a point in [a,b]. We can define a new function on this interval by $F(x) = \int_a^x f \, dx$. We note that F(a)=0. The continuity and differentiability of F on the interval are the subjects of the following theorem.

Theorem 16.

Let f be Riemann integrable on the interval [a,b] and define a function F on [a,b] by

$$F(x) = \int_a^x f(t)\, dt.$$

Then F is continuous on [a,b]. Moreover, if f is continuous at a point x_0 in [a,b], then F is differentiable at that point with $F'(x_0) = f(x_0)$. If f is continuous on the entire interval [a,b], then F is differentiable on the interval with $F' = f$.

We write $\int f(t)\, dt$ rather than $\int f\, dx$ or $\int f(x)\, dx$ to avoid confusion with the variable upper limit of integration. The variables over which we integrate are only dummy variables.

Thus the function defined by the theorem need not be differentiable. A simple illustration of this is given by the function f defined on [0,1] by

$$f(x) = \begin{cases} 1 & \tfrac{1}{2} \le x \le 1 \\ 0 & 0 \le x < \tfrac{1}{2} \end{cases}.$$

In this case we have

$$F(x) = \int_0^x f(t)\, dt = \begin{cases} x - \tfrac{1}{2} & \tfrac{1}{2} \le x \le 1 \\ 0 & 0 \le x < \tfrac{1}{2} \end{cases}.$$

Hence F is continuous on the interval but fails to be differentiable precisely at the point where f is discontinuous, namely at $x = \frac{1}{2}$.

On the other hand, the continuity of f at a point, while a sufficient condition for the differentiability of F, is not a necessary condition. For define F on [0,1] by $F(x) = \int_0^x f(t)\, dt$, where f is a function we have used above,

$$f(x) = \begin{cases} 1/q & x = p/q,\ x\ \text{rational in lowest terms} \\ 0 & x\ \text{irrational} \end{cases}.$$

In this case, $F(x) = 0$ for all x in [0,1], so that $F'(x) = 0$ on the entire interval, both at the irrational points where f is continuous, and at the rational points where f is discontinuous. We note, in this example, that $F'(x_0) = f(x_0)$ only at the irrational points in the interval (and at $x_0 = 0$).

The fundamental theorem of calculus, theorem 15, holds with a variable limit of integration as well as with a fixed limit but, as above, we caution the reader to avoid an often made error. If $F' = f$ on an interval, it does not follow that

$$F(x) - F(a) = \int_a^x f(t)\, dt = \int_a^x F'(t)\, dt.$$

The integral need not be defined as the second example after theorem 15 illustrates.

3. The Necessary and Sufficient Condition for Riemann Integrability.

In section 1 we looked at some classes of functions that are integrable, but we gave no characterization of the general class of integrable functions. To do so is the purpose of this section. However, before we do this, we must introduce a new concept, one generally not encountered in a first course in calculus. This is the concept of a set of measure zero, sometimes referred to as a null set. (Distinguish a null set in this sense from the empty set $\emptyset$.)

In order to define a null set, we must jump a bit ahead of ourselves since we have yet to consider infinite series. But we can make the idea intuitively plausible. Given any set S on the real line, we say the family of open intervals $\{I_\mu\}$ forms a cover for S if $S \subset \bigcup_{\mu=1}^{\infty} I_\mu$. For example if $S = [0,1]$, the open sets $(-\frac{1}{2},\frac{1}{2})$, $(1/4,3/4)$, $(1/2,3/2)$ form an open cover for S. Again, if $S = (0,1]$ and the family of open intervals is $(1/n,2)$, $n = 1,2,\ldots$, then this family of open intervals forms a cover for S but, unlike the first example, no finite collection of this family covers S.

If $I = (a,b)$ is an open interval, we define the length of I to be b-a and, for convenience, we will denote the length by ΔI.

Consider an arbitrary set S on the real line and let $\{I_\mu\}$ be an open cover for S, $S \subset \bigcup_{\mu=1}^{\infty} I_\mu$. We say S is a null set if, given any $\varepsilon > 0$, there exists an open cover for S such that $\sum_{\mu=1}^{\infty} \Delta I_\mu < \varepsilon$ In other words, the sum of the lengths of the open intervals that cover S can be made as small as desired.

For example, consider the set of all rational numbers in the interval [0,1] arranged in some order, $\{a_1, a_2, \ldots\}$. Cover a_1 by an open interval of length $\varepsilon/4$; cover a_2 by an interval of length $\varepsilon/8$; and, in general, cover a_n by an interval of length $\varepsilon/2^{n+1}$. Then $\sum_{\mu=1}^{\infty} \Delta I_\mu = \sum_{n=1}^{\infty} \varepsilon/2^{n+1} = \varepsilon/2 < \varepsilon$. (We shall see this when we consider the geometric series in chapter 6.) Thus the set of rationals in the unit interval is an example of a null set.

We are now in a position to give the necessary and sufficient condition for a function to be Riemann integrable. Continuity on an interval [a,b] is, as we have seen, sufficient but not necessary. Bounded-ness on the interval is necessary but not sufficient. What is necessary and sufficient is given by the following theorem.

Theorem 17.

Let f be a bounded function defined on an interval [a,b]. Then f is Riemann integrable if and only if the set of discontinuities of f on [a,b] is a null set.

Thus, if a bounded function is to be Riemann integrable, it must be continuous almost everywhere, that is, at every point of the interval save, perhaps, points which form a set of measure zero. In fact, this must be the case for every function that is Riemann integrable. And if a bounded function is continuous almost everywhere, it is Riemann inte-grable.

The reader should now be able to see how the function in the example after theorem 12 is inte-grable in spite of having an infinite number of points of discontinuity.

V. TAYLOR'S THEOREM.

The purpose of this short chapter is to consider two topics, polynomial approximations for functions and indeterminate forms. We start with the former.

1. Taylor's Theorem with Remainder.

The reader is already familiar with many different functions, the most simple ones of which are, undoubtedly, the polynomials. For example, let p be the function defined on $[a,b]$ by

$$p(x) = 5x^3 + 3x^2 - x + 4.$$

This is a polynomial function of degree three. Now let us consider a polynomial of degree n defined by

$$P(x) = \sum_{k=0}^{n} a_k(x-x_o)^k = a_o + a_1(x-x_o) + \ldots + a_n(x-x_o)^n,$$

where x_o can be, but need not be, zero. It seems reasonable to assume that there is some connection between the coefficients a_k and the function P. Indeed there is, and the connection is rather simple. We clearly have $a_o = P(x_o)$. If we differentiate P and evaluate the derivative at $x = x_o$, we find $P'(x_o) = a_1$. Continuing this process, it is not difficult to show that

$$a_k = P^{(k)}(x_0)/k!$$

Hence, the polynomial function can be written as

$$P(x) = P(x_0) + P'(x_0)/1! \ (x-x_0) + P''(x_0)/2! \ (x-x_0)^2$$

$$+ \ldots + P^{(n)}(x_0)/n! \ (x-x_0)^n.$$

Now consider a function f which is n times differentiable at $x = x_0$ defined on some interval. We
are naturally led to ask the question how is

$$f(x_0) + f'(x_0)/1! \ (x-x_0) + \ldots + f^{(n)}(x_0)/n! \ (x-x_0)^n \ (*)$$

related to f(x). Since, in general, f will not be a
polynomial, we would not expect f to be equal to the
polynomial (*). Yet we should expect some connection
between this polynomial function and f. This relationship is given by Taylor's formula.

Theorem 1. (Taylor's Formula)

Let f be n times continuously differentiable
on the interval $[x_0, x]$. Let the n+1 st derivative
of f exist and be integrable on this interval. Then

$$f(x) = T_{n,x_0}(x) + R_{n,x_0}(x)$$

where

$$T_{n,x_o}(x) = f(x_o) + f'(x_o)/! \ (x-x_o) + \ldots$$

$$+ \ f^{(n)}(x_o)/n! \ (x-x_o)^n$$

$$= \sum_{k=0}^{n} f^{(k)}(x_o)/k! \ (x-x_o)^k \ ;$$

and $R_{n,x_o}(x) = 1/n! \int_{x_o}^{x} (x-t)^n f^{(n+1)}(t) \ dt$.

Moreover, if $P(x)$ is any other polynomial which equals f up to order n at x_o, then $P = T_{n,x_o}$.

By the last statement of the theorem we mean

$$\lim_{x \to x_o} \frac{f(x) - P(x)}{(x-x_o)^n} = 0.$$

Thus the n th order "Taylor polynomial" is the unique polynomial which agrees with f up to the n th order at x_o. The requirement that the n+1 st derivative both exist and be integrable is essential in this formulation of Taylor's theorem. The reason is, as we have seen in chapter 4, that there are functions possessing a primitive on a closed interval which, nevertheless, fail to be integrable on the interval.

Theorem 1 is referred to as Taylor's formula with integral remainder. The theorem is often formulated under the assumption that f is n+1 times continuously differentiable, but this assumption is stronger than is necessary.

If $f^{(n+1)}$ is continuous on $[x_0, x]$, it is easy to get an upper and lower bound on the remainder term. The reader should have no difficulty in citing the theorem to justify this statement and in finding the upper and lower bounds. The remainder term is basically a measure of how well the n th Taylor polynomial approximates the function f in the neighborhood of x_0. We might note that we assumed $x > x_0$ in the statement of the theorem. If $x < x_0$, the theorem remains true with appropriate modifications.

The integral form of the remainder is not the only form or, for that matter, the most convenient one. We will obtain other forms by stating a theorem not generally found in elementary calculus texts.

Theorem 2.[*]

Let f be n times continuously differentiable on the interval $[x_0, b]$ and let n+1 st derivative of f exist on (x_0, b). Moreover, let ϕ and ψ be functions continuous on $[x_0, b]$ and differentiable on (x_0, b).

L.M. Blumenthal, Am. Math Monthly, 33 (1926), 424-426.

Suppose that for all x and y in (x_0, b) with $x_0 < y < x$

the determinant

$$\begin{vmatrix} \phi'(y) & \psi'(y) \\ \phi(x) & \psi(x) \end{vmatrix} \neq 0.$$

Then for all x in $(x_0, b]$, there exists a ξ in (x_0, x)

such that

$$f(x) = \sum_{k=0}^{n} f^{(k)}(x_0)/k! + R_{n, x_0}(x)$$

$$= T_{n, x_0}(x) + R_{n, x_0}(x),$$

where

$$R_{n, x_0}(x) = \frac{-\begin{vmatrix} \phi(x_0) & \psi(x_0) \\ \phi(x) & \psi(x) \end{vmatrix}}{\begin{vmatrix} \phi'(\xi) & \psi'(\xi) \\ \phi(x) & \psi(x) \end{vmatrix}} \; f^{(n+1)}(\xi)/n! \; (x-\xi)^n.$$

We refer the reader to the article cited above

for the proof of this theorem. Theorem 2 is stronger

than theorem 1 in that the first theorem assumes the

Riemann integrability of $f^{(n+1)}$, whereas the second

assumes only the existence of the n+1 st derivative

on the interval.

It is clear that we can now generate any number of remainder terms for Taylor's formula. Not all, however, are of equal importance. We will merely give a few of the more common ones.

If we take ψ to be any nonzero constant, we get

$$R_{n,x_o}(x) = \frac{\phi(x) - \phi(x_o)}{\phi'(\xi)} \; f^{(n+1)}(\xi)/n! \; (x-\xi)^n.$$

This is the Schömilch form of the remainder.

If we now set $\phi(t) = (x-t)^p$ in the Schömilch remainder, where $1 \le p \le n+1$, we obtain the Roche form of the remainder,

$$R_{n,x_o}(x) = f^{(n+1)}(\xi)/pn! \; (x-x_o)^p(x-\xi)^{n-p+1}.$$

Again, setting $p = 1$ and $p = n+1$ in the Roche remainder, we obtain respectively, the Cauchy form for the remainder,

$$R_{n,x_o}(x) = f^{(n+1)}(\xi)/n! \; (x-x_o) \; (x-\xi)^n$$

and the Lagrange form for the remainder,

$$R_{n,x_o}(x) = f^{(n+1)}(\xi)/(n+1)! \; (x-x_o)^{n+1} \; .$$

These last two forms, along with the integral form of the remainder, are the ones most frequently encountered.

If the reader would write out Taylor's formula with $n = 0$ using the Lagrange form of the remainder, he would find the expression for the mean value theorem. In this sense, Taylor's formula gives a generalization of the mean value theorem.

The results above are often written in a different fashion. For example, if we write $x = x_o + h$ and $\xi = x_o + \theta h$, where $0 < \theta < 1$, the Roche form of the remainder becomes

$$R_{n,x_o}(x) = f^{(n+1)}(x_o + \theta h)/pn!\ h^{n+1}(1-\theta)^{n-p+1}$$

and Taylor's formula can be expressed as

$$f(x_o + h) = \sum_{k=0}^{n} f^{(k)}(x_o)/k!\ h^k$$
$$+ f^{(n+1)}(x_o + \theta h)/pn!\ h^{n+1}\ (1-\theta)^{n-p+1}.$$

Similar expressions occur with the Cauchy and the Lagrange forms of the remainder.

This completes our discussion of Taylor's formula for functions of one variable. However, we will have to return to consider the remainders further when we take up the discussion of Taylor series.

2. L'Hôpital's Rule.

The reader has already seen the elementary arithmetic rules for limits. For instance, the limit of a quotient equals the quotient of the limits of the numerator and the denominator provided these limits exist and the limit of the denominator is unequal to zero, that is,

$$\lim_{x \to a} f(x)/g(x) = \lim_{x \to a} f(x)/\lim_{x \to a} g(x) \text{ when } \lim_{x \to a} g(x) \neq 0.$$

However, it may happen that the limits of both f and g vanish as x approaches a. Then the quotient of the limits is of the form 0/0. Such an expression is called an indeterminate form, and it is such forms we wish to consider now.

As an example, what can be said of the limit

$$\lim_{x \to 0} (\sin x)/x \ ?$$

The reader may have seen a geometric argument showing that this limit is one. Another line of approach, one of more general applicability, would be to use Taylor's theorem with remainder. In the neighborhood of the origin we would find

$$\sin x = x - \cos(\xi)/3! \ x^3, \quad 0 < \xi < x.$$

105.

Hence,

$$\lim_{x \to 0} \sin x/x = \lim_{x \to 0} \{x - \cos \xi/3! \ x^3\} /x$$

$$= 1 - \lim_{x \to 0} \cos \xi/3! \ x^2 = 1.$$

But there is another way to treat indeterminate forms which is generally more efficient and easier to apply than the above method. This method is known as L'Hôpital's rule.

Theorem 3. (L'Hôpital's rule for indeterminate forms of the type 0/0)

Let f and g be functions whose limits go to zero as $x \to a^+$. Moreover, suppose that f and g are differentiable at each point of an open interval (a,b) and that $g'(x) \neq 0$ for each x in (a,b). Then if the limit $\lim_{x \to a^+} f'(x)/g'(x)$ exists and has the value A, where A is a real number or $\pm \infty$, the limit $\lim_{x \to a^+} f(x)/g(x)$ also exists and has the value A.

We first remark that similar remarks hold for left-handed limits and for two sided limits. Furthermore, the theorem is valid for $x \to \infty$ or $x \to -\infty$. In this case we would require that the functions be differentiable on a semi-infinite interval.

We also note that the theorem does not require that f'(a) or g'(a) exist. It is only the limit of the quotient that must exist. (Recall the definition of the limit!)

Certainly the application of the theorem is simplicity itself. But the reader should be warned that several applications of l'Hôpital's rule may be necessary before the answer is obtained. Again, some care is necessary. The form must be indeterminate! For example, consider the function

$$f(x) = \frac{5x^2 - 4x - 1}{x^3 - 1} \qquad \text{as x approaches one.}$$

A mechanical application of the theorem gives

$$\lim_{x \to 1} \frac{5x^2 - 4x - 1}{x^3 - 1} = \lim_{x \to 1} \frac{10x - 4}{3x^2} = \lim_{x \to 1} \frac{10}{6x} = 5/3.$$

The first step is correct but the following step is not since the second expression is not an indeterminate form. The correct limit is had from the second expression. The answer is 2.

Moreover, l'Hôpital's rule is not infallible. The theorem may not be applicable, yet the limit may exist. For instance, consider the form

$$\lim_{x \to 0} \frac{x^2 \sin(1/x)}{\sin x} .$$

A straightforward application of l'Hôpital's rule gives

$$\lim_{x \to 0} \frac{2x \sin(1/x) - \cos(1/x)}{\cos x}$$

The limit of the numerator fails to exist. However, if we write the original limit as a product of limits,

$$\lim_{x \to 0} x/\sin x \quad \lim_{x \to 0} x \sin(1/x),$$

we see the limit of the original expression does exist and is equal to zero.(L'Hôpital's rule was applied to the first factor in the product of the limits.) Hence,we conclude that theorem 3 gives a sufficient condition for the limit of the quotient to exist, but the condition is not necessary.

As another example, consider

$$\lim_{x \to \infty} x e^{-x} = \lim_{x \to \infty} \frac{e^{-x}}{1/x} \quad .$$ Using theorem 3 on this

indeterminate form, we get $\lim_{x \to \infty} \frac{-e^{-x}}{-1/x^2} = \lim_{x \to \infty} x^2 e^{-x}$,

so that the last state of the man is worse than the first. Repeated applications of the rule would lead us deeper and deeper into frustration. But this example suggests a second form of l'Hôpital's rule.

Theorem 4. (L'Hôpital's rule for indeterminate forms

of the type ∞/∞)

Let f and g be functions whose limits go to $+\infty$ or $-\infty$ as $x \to a^+$. Moreover, suppose that f and g aredifferentiable at each point of an open interval (a,b) and that $g'(x) \neq 0$ for each x in (a,b). Then if the limit $\lim_{x \to a} f'(x)/g'(x)$ exists and has the value A, where A is a real number or $\pm\infty$, the limit $\lim_{x \to a^+} f(x)/g(x)$ also exists and has the value A.

The same remarks that follow theorem 3 are likewise in order for theorem 4.

Returning to the example that preceded this theorem, write $\lim_{x \to \infty} xe^{-x} = \lim_{x \to \infty} x/e^x$. This expression now is an indeterminate form of the type ∞/∞ and a straightforward application of theorem 4 shows that the limit is zero.

Again, if we consider $\lim_{x \to \infty} \dfrac{x-\sin x}{x}$, a form of the type ∞/∞, and if we apply theorem 4, we obtain $\lim_{x \to \infty}$ 1- cosx/1. This limit does not exist! But if we write the original expression as $\lim_{x \to \infty}$ 1 - $\lim_{x \to \infty}$ sin x/x, we see that the limit does exist and is equal to one.

Hence, in this case too, the theorem states a sufficient condition for the limit of the indeterminate form to exist, not a necessary condition.

There are other types of indeterminate forms. For example, one of the type $0 \cdot \infty$ can be put into the form $0/0$ or ∞/∞. It can then be treated by using theorem 3 or theorem 4. We did this in the example following theorem 4.

Indeterminate forms of the type $\infty - \infty$ are generally brought into the type $0/0$ or ∞/∞ by algebraic manipulation. For example,

$$\lim_{x \to 0} \{e^{-x}/x - 1/(e^x - 1)\} = \lim_{x \to 0} \frac{1 - e^{-x} - x}{x(e^x - 1)} \quad ,$$

and a double application of theorem 3 shows that the limit is $-1/2$. Indeed, in cases of this type, it is often true that the algebraic operations will reduce the form to an expression that is not indeterminate.

The indeterminate expressions 1^{∞}, ∞^0, and 0^0 are each treated in the following fashion. Consider $\lim_{x \to a^+} \{f(x)\}^{g(x)}$, where as $x \to a^+$,

1. $f(x) \to 1$ and $g(x) \to \infty$;
2. $f(x) \to \infty$ and $g(x) \to 0$;
3. $f(x) \to 0$ and $g(x) \to 0$.

By virtue of the continuity of the logarithm, we have

$$\ln\{\lim_{x\to a^+} [f(x)]^{g(x)}\} = \lim_{x\to a^+}\{\ln (f(x))^{g(x)}\}$$

$$= \lim_{x\to a^+}\{g(x)\cdot\ln f(x)\} \ .$$

Thus, the limit is of the type $\pm\infty\cdot 0$. We can reduce this to one of the standard forms and then apply l'Hôpital's rule. When the limit of the logarithm is obtained, we then exponentiate to obtain the limit of the original expression.

We end this chapter with a remark. Beginning students are often unconvinced that the expressions 0^0, 1^∞, and ∞^0 are really indeterminate. Isn't anything to the zero power one? Isn't one to any power one? This opinion is reinforced by most of the examples the student may find in calculus texts. But these forms are, indeed, indeterminate and can have limits unequal to one.

For example, $\lim_{x\to\infty} \{e^{-ax} + e^{-bx}\}^{1/x}$, where $a,b > 0$ and $a < b$, is of the form 0^0 and the limit is e^{-a}.

Again, $\lim_{x \to 1} x^{1/(1-x)}$ is of the form 1^∞ and the limit of this form is e^{-1}.

Finally, $\lim_{x \to \infty} (e^{ax} + e^{bx})^{1/x}$, where $a, b > 0$ and $a < b$, is of the form ∞^0 and the limit is e^b.

It would be well for the reader to verify these results and to construct others of a similar type where the limit is not one.

VI. NUMERICAL SEQUENCES AND SERIES.

1. Sequences.

In this chapter we shall consider the principal properties and theorems which apply to sequences and infinite series of real numbers. In the following chapter, we shall treat the somewhat more difficult topic of series and sequences of functions.

We start by defining what is meant by a sequence. Quite simply, a sequence is a function, with the word function having exactly the same meaning as it had earlier in this book. What distinguishes a sequence from the functions we discussed in previous chapters is its domain. The domain of a sequence is not the real line or some interval of the real line, but the positive integers. At times it may be more convenient to take the domain to be the nonnegative integers. Actually, if n_o is any positive integer, we could just as well take the domain of the sequence to the integers $n \geq n_o$.

A sequence in this chapter maps the integers into a subset of the reals, symbolically $f: Z^+ \to R$. To be consistent with our earlier notation, we should write $f(n)$ or $a(n)$ for the value of the function f or a at the point n. However, in the case of sequences, it is almost universal practice to write the value of the function at n as f_n or a_n. Again, it is almost universal practice to denote a sequence by its values as $\{f_n\}$ or $\{a_n\}$.

Some examples would be:

$\{1/n^2\}$ = 1, 1/4, 1/9, ...

$\{\sin(n\pi/2)\}$ = sin π/2, sin π, sin 3π/2, ...

$\{(-1)^n\}$ = -1, 1, -1, 1, ...

$\{\pi\}$ = π, π, π, ...

Given a sequence, we naturally ask what value, if any, the sequence has as $n \to \infty$. This is a question of limits and convergence.

Definition.

The sequence $\{a_n\}$ converges to a (or has a limit a), written $\lim_{n \to \infty} a_n = a$, if for all ε > 0, there is an integer N such that $|a - a_n| < \varepsilon$ for all $n \geq N$

If a sequence does not converge, then it diverges. For example, the sequence $\{n^2\}$ diverges. However, divergence does not necessarily imply that the n th term of the sequence grows arbitrarily

114.

large as n gets large. Such sequences do, indeed, diverge, but the sequence $\{(-1)^n\}$ also diverges. It has no limit!

Hence the range of a sequence may be bounded, even finite, yet the sequence may fail to converge. By the range of the sequence we mean, of course, the values the sequence takes on. This set of values may be finite or infinite. A sequence is bounded if there is a real number M such that $|a_n| \le M$ for all n. Be careful to distinguish the notion of range of a sequence which may be a finite or an infinite set, and the notion of boundedness of the range. Not every bounded sequence converges, but if the sequence does converge, it is bounded.

The first characteristic of a convergent sequence is important enough to be stated as a theorem.

Theorem 1.

If a sequence $\{a_n\}$ converges to a limit, then that limit is unique.

It is obvious, however, that two or, for that matter, any number of sequences can converge to the same limit. For example, the sequences $\{1/2n\}$ and $\{1/(2n-1)\}$ both have the same limit.

Theorem 2.

Let $\lim\limits_{n\to\infty} a_n = a$ and $\lim\limits_{n\to\infty} b_n = b$. Then

1. $\lim\limits_{n\to\infty} (a_n + b_n) = \lim\limits_{n\to\infty} a_n + \lim\limits_{n\to\infty} b_n = a + b$;

2. $\lim\limits_{n\to\infty} a_n b_n = \lim\limits_{n\to\infty} a_n \cdot \lim\limits_{n\to\infty} b_n = a \cdot b$;

3. $\lim\limits_{n\to\infty} |a_n| = |a|$;

4. if $b_n \neq 0$ for all n and if $b \neq 0$,

   $\lim\limits_{n\to\infty} a_n / b_n = \lim\limits_{n\to\infty} a_n / \lim\limits_{n\to\infty} b_n = a/b$;

5. $\lim\limits_{n\to\infty} k a_n = k \lim\limits_{n\to\infty} a_n = k_a$

A few comments on various parts of this theorem
are in order. The sum of a convergent and divergent
sequence diverges, but the sum of two divergent sequences may converge. For instance, take the sequences $\{a_n\}$ and $\{b_n\}$, where $a_n = (-1)^n$ and
$b_n = (-1)^{n+1}$. Both sequences diverge, but their sum
is the constant sequence $\{0\}$, which certainly converges. The same example shows that the product and
the quotient of two divergent sequences may converge.

However, with the product and the quotient, we

may have the product and the quotient of a convergent sequence and a divergent sequence converging.
For example, let $\{a_n\}$ be the sequence $\{1/n\}$ and
let $\{b_n\}$ be the sequence $\{(-1)^n\}$. Moreover, in
(4), $\lim\limits_{n\to\infty} a_n/b_n$ may well exist even if $\lim\limits_{n\to\infty} b_n = 0$,
as with the sequences $\{a_n\} = \{1/n^2\}$ and $\{b_n\} = \{1/n\}$

Finally, the convergence of $\{a_n\}$ implies the
convergence of $\{|a_n|\}$ but not vice versa as the
sequence $\{(-1)^n\}$ shows.

If two sequences $\{a_n\}$ and $\{b_n\}$ are such that
$a_n \leq b_n$ for all n, the sequence $\{b_n\}$ is said to
dominate the sequence $\{a_n\}$. If both sequences converge, say to a and to b respectively, it is
natural to conclude that $a \leq b$. This is, indeed, the
case. However, strict inequality $a_n < b_n$ for all n,
need not imply $a < b$. The reader should have little
difficulty in formulating examples to illustrate
this fact.

But we must be cautious on the question of a
dominating sequence. As we shall see, a dominating
series can be used to test the convergence of
another series. This is not true with sequences.
Take the sequences $\{a_n\} = \{(-1)^n\}$ and

and $\{b_n\} = \{(n+1)/n.\}$ Then $a_n < b_n$ and $\{b_n\}$ converges; but $\{a_n\}$ diverges. In all of the examples above, it was rather simple to determine whether the sequences converge or diverge. In general, however, this is not the case. It may be extremely difficult to determine the convergence or divergence of a given sequence and, as the previous example illustrates, there is no simple comparison test to aid in the task. But there is one type of sequence for which the situation is far more agreeable.

Definition.

A sequence $\{a_n\}$ is called a monotonic increasing (decreasing) sequence if $a_n \leq a_{n+1}$ ($a_n \geq a_{n+1}$) for all n.

Theorem 3.

Let $\{a_n\}$ be bounded above and increasing (bounded below and decreasing). Then $\{a_n\}$ converges and $\lim_{n\to\infty} a_n = \sup_n a_n$ ($\inf_n a_n$).

We have already seen that a convergent sequence is bounded. Hence a monotonic sequence converges if and only if it is bounded. This theorem is useful

118.

since it enables us to determine the convergence of certain types of sequences even if we do not know their limits.

A nontrivial example of the usefulness of this theorem is the following. Consider the sequence $\{a_n\}$,

$$a_n = 1 + 1/2 + 1/3 + \ldots + 1/n - \ln n.$$

It is not difficult to show that each term of this sequence is positive and that the sequence is mono-tonic decreasing. Hence, this sequence is bounded below by zero. By theorem 3, then, the sequence converges. Of course, the theorem does not give us the limit. In fact, the limit in this case,

$$\gamma = \lim_{n \to \infty} a_n = 0.57721566\ldots,$$

is known as Euler's constant. It is not known whether γ is a rational or irrational number.

Let us return to the sequence $\{(-1)^n\}$. As we know, this sequence diverges. But if we were to con-sider the sequence formed from all the odd (or even) terms of this sequence, this situation is altered. Hence we are led to the following definition.

Definition.

Let $\{a_n\}$ be a given sequence and let $\{n_k\}$ be a

sequence of positive integers such that
$n_1 < n_2 < n_3 < \ldots$. The sequence $\{a_{n_k}\}$ is called a
subsequence of the sequence $\{a_n\}$.

For example if $\{a_n\}$ is $\{1/n\} = 1, 1/2, 1/3,\ldots,$
then $\{a_{2n-1}\} = \{1/(2n-1)\} = 1, 1/3, 1/5, \ldots$ is
a subsequence of $\{a_n\}$. The sequence
$1, 1/2, 1/6, 1/27, 1/28, 1/113, \ldots$ would also be a
subsequence of $\{a_n\}$, whereas $1, 1/2, 1/6, 1/3, 1/8,$
$\ldots$ would not be. This is the import of
$n_1 < n_2 < n_3 < \ldots$.

Now it is clear from the definition that if a
given sequence converges, every subsequence of it
will converge and to the same limit. But a given
sequence can well diverge and yet have one or more
convergent subsequences. In fact, the sequence can
be not only divergent but also unbounded and still
have a convergent subsequence. For example, consider
the sequence $\{a_n\}$ whose values are given by

$$a_n = \begin{cases} n & n \text{ odd} \\ 1/n & n \text{ even} \end{cases} .$$

The even terms give a convergent subsequence of this
unbounded sequence.

A given sequence, convergent or divergent, can have two or more subsequences which converge to the same (subsequential) limit. But if a sequence has subsequences which converge to different subsequential limits, we can be assured that the original sequence is divergent.

In order to pursue our consideration of sequences further, we should first state an important result from set theory.

Definition.

Let S be a set of real numbers. Then a is a limit point of S if and only if, for each $\varepsilon > 0$, the open interval $(a-\varepsilon, a+\varepsilon)$ contains a point of S distinct from a.

It is important to distinguish the idea of limit point of a set from limit of a sequence. If a sequence has a limit, that limit is unique. But if we consider the range of values of a sequence as a set of real numbers, even a divergent sequence may have more than one limit point.

Again, a limit point need not belong to the set. For example, 0 is a limit point of the set $\{1, 1/2, 1/3, \ldots\}$, but it does not belong to the set. In this example it is clear that every ε-neighborhood of 0 contains infinitely many members of the set. A little thought should convince the reader that, in general, this is true of all limit points of a set. That is to say, if a is a limit point of the set S, every ε-neighborhood of a contains infinitely many members of S distinct from a.

In the example in the previous paragraph, you note that the limit point 0 is the greatest lower bound of the set. It can be shown that if m is the greatest lower bound of a set S and if $m \notin S$, then m is a limit point of S. However, if $m \in S$, it need not be a limit point. Consider the set {n: n integer, n $\geq$ 0}. The greatest lower bound of this set is 0, but 0 is not a limit point of this set. Analogous remarks hold for the least upper bound.

It should be noted that some authors use the term accumulation point or cluster point rather than limit point.

Theorem 5.

If S is a set of real numbers with a limit point a, then there is a sequence $\{a_n\}$ converging to a, where a_n is in S for each n and each of the values

a_1, a_2, a_3, ... is distinct.

The following important result is a consequence of the previous two theorems.

Theorem 6.

If $\{a_n\}$ is a bounded sequence, then $\{a_n\}$ contains a convergent subsequence.

We saw that an unbounded sequence can have, but need not have, a convergent subsequence. But by virtue of this theorem, if the sequence is bounded, it must have a convergent subsequence. This is true whether the range of the sequence is a finite or infinite set.

Let us consider in more detail the range of a sequence. If the range of a sequence $\{a_n\}$ is a finite set, then the only way that the sequence can converge is that the values a_n all be the same for all $n \geq N$, where N is some fixed integer. If the range of the sequence is an infinite set, the situation is more involved.

Theorem 7.

If $\{a_n\}$ is a convergent sequence and if S, the set of distinct points of the range of $\{a_n\}$, is an infinite set, then S has only one limit point and this point is the limit of the sequence.

However, the set of distinct points of the range of $\{a_n\}$ can be an infinite set with only one limit point, but the sequence need not converge. For example, let

$$a_n = \begin{cases} 1 & n \text{ odd} \\ 1/n & n \text{ even} \end{cases} .$$

But we do have a quasi converse to theorem 7.

Theorem 8.

Let $\{a_n\}$ be a bounded sequence such that the set of points $S = \{a_1, a_2, a_3, \ldots\}$ are all distinct. If S has only one limit point, a, then $\{a_n\}$ converges to the limit a.

The condition of this theorem that fails to hold in the example after theorem 7 is that the a_1, a_2, a_3, $\ldots$ are all to be distinct. Again, it is essential that the sequence be bounded. We have already seen an example of a divergent unbounded sequence such that the set of points a_1, a_2, a_3, $\ldots$ are distinct and possess only one limit point.

The reader should have noted that, with the exception of monotonic sequences, it was necessary to know the limit of a sequence in order to test for convergence. This is, of course, a highly unsatis-factory situation. What if the limit is unknown? Or what if we are interested only in the fact of con-vergence or divergence of a sequence and not in the value to which the sequence may converge? This sit-uation can be resolved by using Cauchy's criterion.

Definition.

A sequence $\{a_n\}$ is called Cauchy convergent if, for all $\varepsilon > 0$, there is an integer N such that $|a_n - a_m| < \varepsilon$ for all $m, n \geq N$; (or equivalently, $|a_{n+k} - a_n| < \varepsilon$ for all $n \geq N$ and all $k > 0$).

In the equivalent formulation it is important to emphasize for all k.   For consider the sequence $\{a_n\} = \{n^{\frac{1}{2}}\}$. Then $|a_{n+k} - a_n| = (n+k)^{\frac{1}{2}} - (n)^{\frac{1}{2}} < k/2n^{\frac{1}{2}}$. Now for any fixed k, we can find an N such that $k/2N^{\frac{1}{2}} < \varepsilon$, and so $|a_{n+k} - a_n| < \varepsilon$ for all $n \geq N$. However, if this situation is to hold for all $k > 0$, the situation is changed. The right hand side of the inequality cannot be made arbitrarily small. This sequence is, in fact, unbounded and hence diverges.

Theorem 9.

A sequence of real numbers converges if and only if it is Cauchy convergent.

By virtue of this theorem, we can now test a sequence for convergence without knowing its limit.

The fact that every Cauchy sequence of real numbers converges states a fundamental property of the real number system, namely its completeness. Not every number system is complete. For example, suppose our system was the rational numbers. The reader can easily construct a sequence of rational numbers which converges to $2^{\frac{1}{2}}$ for instance. Such a sequence would be a Cauchy sequence but it would not be convergent in the sense that its limit, the irrational number $2^{\frac{1}{2}}$, does not belong to the system. Thus the rational number system, unlike the real number system, is not complete.

We end our discussion of Cauchy sequences by noting that, assuming the usual properties of the real number system, theorem 3, theorem 4, and theorem 9 of this chapter, as well as theorem 9 of chapter are all equivalent. That is to say, each theorem implies and is implied by the other. Hence, in proving a result or solving a problem that requires any one

126.

of these theorems, it is simply a matter of convenience as to which theorem to use.

Let us return to the sequence $\{a_n\} = \{(-1)^n\}$.
Its range is finite, +1 and -1. Consider an ε-neigh-
borhood of 1, that is the open set $(1-\varepsilon, 1+\varepsilon)$ where
$\varepsilon > 0$. Infinitely many a_n belong to this neighborhood
(and similarly with any ε-neighborhood of -1). The
points +1 and -1 are called cluster points of the
sequence.

Definition.

A point a is called a cluster point of the
sequence $\{a_n\}$ if every open interval centered at a
contains a_n for infinitely many n.

A sequence can have any number of cluster
points but,if the sequence converges to a limit, say
a, then a is the only cluster point. In this event,
any open interval centered at a contains a_n for all
but a finite number of n. This is different from
saying "contains a_n for infinitely many n".

We should distinguish cluster point of a sequence from cluster point of a set (what we called
limit point of a set above). If a is a cluster
(limit) point of a set S, every open interval cen-

tered at a must contain infinitely many members of S distinct from a. If a is a cluster point of a sequence, then a_n need not be distinct from a. In the sequence $\{a_n\} = \{(-1)^n\}$, they are not.

Again, a cluster point of a sequence may, but need not, be a limit point of the range of the sequence. In the example above, +1 and -1 are not limit points of the ranges of the sequence $\{(-1)^n\}$. But if a point a is a limit point of the range of a sequence $\{a_n\}$, then it is a cluster point of the sequence.

Now we have seen that every bounded sequence has at least one convergent subsequence. The limit of that subsequence will be a cluster point of the original sequence. Consider the set S of all cluster points of a given sequence $\{a_n\}$. Since the sequence is bounded, the set S will be bounded and hence, will have a least upper bound and a greatest lower bound. The least upper bound of this set of cluster points is called the limit superior of the sequence and is denoted by $\varlimsup_{n\to\infty} a_n$ or $\limsup_{n\to\infty} a_n$. Similarly, the greatest lower bound of the set of cluster points of the sequence is called the limit inferior of the sequence and is written as

128.

$\varliminf_{n\to\infty} a_n$ or $\lim\inf_{n\to\infty} a_n$.

It is clear that $\varliminf a_n \leq \varlimsup a_n$. If the sequence is unbounded above, we have $\varlimsup a_n = \infty$, and if the sequence is unbounded below, we have $\varliminf a_n = -\infty$.

Theorem 10.

The sequence $\{a_n\}$ converges if and only if $\varlimsup a_n = \varliminf a_n = \lim a_n$.

On the other hand, if $\lim_{n\to\infty} a_n = \infty$, we have $\varlimsup_{n\to\infty} a_n = \varliminf_{n\to\infty} a_n = \infty$, with a similar result if $\lim_{n\to\infty} a_n = -\infty$.

A word of caution is in order here. The range of a bounded sequence $\{a_n\}$ has a least upper bound, $\sup_n a_n$, and a greatest lower bound, $\inf_n a_n$. These may coincide with the superior and inferior limits respectively of the sequence. For example, if a_n is a monotonic increasing sequence which is bounded above and converges to a, then $\sup a_n = a$. However, in general, there is no reason to expect these values to be the same. For instance, consider the sequence

$\{a_n\}$, where

$$a_n = \begin{cases} 4 + 1/n & n \text{ even} \\ 1 - 1/n & n \text{ odd} \end{cases} ,$$

that is, $\{0, 4\ 1/2,\ 2/3,\ 4\ 1/4,\ 4/5,\ 4\ 1/6,\ \ldots\}$.

In this case we have $\overline{\lim}\ a_n = 4$ and $\underline{\lim}\ a_n = 1$,

whereas sup $a_n = 4\ 1/2$ and inf $a_n = 0$.

Although the definition of inferior and superior
limits should be clear now, it may be helpful to
state the definition in terms of ε and N.

Definition.

1. The real number a is the limit superior of
the sequence $\{a_n\}$ if and only if

   i) for all $\varepsilon\ >\ 0$, there exists an N such
       that $a_n\ <\ a + \varepsilon$ for all $n \geq N$, and

   ii) for all $\varepsilon\ >\ 0$ and for all N, there exists
       an $n \geq N$ such that $a_n\ > a - \varepsilon$.

2. The real number a is the limit inferior of
the sequence $\{a_n\}$ if and only if

   i) for all $\varepsilon\ >\ 0$, there exists and N such
       that $a_n\ > a - \varepsilon$ for all $n \geq N$, and

   ii) for all $\varepsilon\ >\ 0$ and for all N, there exists
       an $n \geq N$ such that $a_n\ <\ a + \varepsilon$.

The previous example is a good illustration of the definition.

Theorem 11.

Let $\{a_b\}$ and $\{b_n\}$ be two given sequences. Then

1) $\underline{\lim}\ a_n + \underline{\lim}\ b_n \leq \underline{\lim}(a_n+b_n) \leq \overline{\lim}(a_n+b_n)$

$$\leq \overline{\lim}\ a_n + \overline{\lim}\ b_n, \qquad \text{and}$$

2) if $a_n \geq 0$ and $b_n \geq 0$, then

$(\underline{\lim}\ a_n)(\underline{\lim}\ b_n) \leq \underline{\lim}\ a_n b_n \leq \overline{\lim}\ a_n b_n$

$$\leq (\overline{\lim}\ a_n)\,(\overline{\lim}\ b_n)$$

provided the product is defined.

The reader should note that strict inequality can occur. The following example is such that strict inequality holds for each **inequality** in both the sum and product. Let

$\{a_n\} = \{1,\ 2,\ 3,\ 4,\ 1,\ 2,\ 3,\ 4,\ 1,\ \dots\}$ and

$\{b_n\} = \{3,\ 4,\ 1,\ 2,\ 3,\ 4,\ 1,\ 2,\ 3,\ \dots\}$. Then

$\{a_n+b_n\} = \{4,\ 6,\ 4,\ 6,\ 4,\ \dots\}$ and

$\{a_n b_n\} = \{3,\ 8,\ 3,\ 8,\ 3,\ \dots\}$.

Thus, part 1 of the theorem yields the inequality $2 < 4 < 6 < 8$, and part two the inequality $1 < 3 < 8 < 16$.

Again, for the product inequality, it is important that a_n, $b_n \geq 0$, as the sequences

$\{a_n\}$ = $\{0, 1, -1, 0, 1, -1, 0, \ldots\}$ and

$\{b_n\}$ = $\{1, 0, -1, 1, 0, -1, 1, \ldots\}$ illustrate.

We leave the details to the reader.

This concludes our discussion of limit superior and limit inferior for the present. We shall return to these notions later, however, when we consider convergence tests for infinite series.

We want to close this section by giving another definition of continuity, a definition which we are now in position to formulate.

Definition. (The Sequential Definition of Continuity)

Let f be a real valued function defined on an interval I and let $x_0 \in$ I. Then f is continuous at x_0 if and only if $\lim_{n \to \infty} f(x_n) = f(x_0)$ for every sequence $\{x_n\}$ with terms belonging to I which is such that $\lim_{n \to \infty} x_n = x_0$.

II. Series

1. Convergence of Infinite Series.

The meaning of the sum of a finite number of terms, say $\sum_{k=1}^{n} a_k$, is quite clear and the usual algebraic properties of such sums are immediate. The problem at hand is to consider the sum of terms when the number of summands is not finite. As of yet, an expression such as $\sum_{n=1}^{\infty} a_n$ has not been defined. However, our study of sequences in the first part of this chapter provides us with a means of defining the expression $\sum_{n=1}^{\infty} a_n$, a quantity which will be called an infinite series or just a series.

Consider a numerical sequence $\{a_n\}$ and the series generated by it, $\sum_{n=1}^{\infty} a_n$. From the given sequence $\{a_n\}$ we can form a new sequence $\{s_n\}$, the sequence of partial sums, where $s_n = \sum_{k=1}^{n} a_k$. If the sequence of partial sums $\{s_n\}$ converges, we define its limit to be the sum of the infinite series, that is, $\lim_{n \to \infty} s_n = \sum_{n=1}^{\infty} a_n$, and we say that the series converges.

There is a drawback to this definition in that not every series will converge since, as we have seen, not every sequence converges. This is not too disturbing, perhaps, when the initial sequence is unbounded, say $\{a_n\} = \{n\}$. In this case we have $\lim_{n\to\infty} s_n = 1 + 2 + 3 + \ldots = \infty$. However, in the event that the original sequence is bounded, we may feel less than satisfied with the definition. For example, take the bounded sequence $\{a_n\} = \{(-1)^{n+1}\}$. The sequence of partial sums $\{s_n\} = \{1, 0, 1, 0, 1, \ldots\}$ has no limit and hence, the series fails to converge. If we write out the series, we get $\sum_{n=1}^{\infty} (-1)^{n+1} = $ 1-1+1-1+1=1+ .... According as how we group the terms, this sum is 0 or 1.

Why not just define $\sum_{n=1}^{\infty} (-1)^{n+1}$ to be the average 1/2? The reader thinking along these lines need not be ashamed. He is in the company of a number of famous mathematicians. But the fact remains that with our definition of convergence for a series, -- and this is the definition we shall use throughout this book,-- this series fails to converge.

We should remark that there are extended definitions of convergence wherein the above series would converge and, indeed, to the value 1/2. Moreover, if a given series converges according to our definition, it would converge with the extended definition and to the same value. But these matters go beyond the scope of what we wish to discuss in this chapter. One further remark: we have written a series as $\sum\limits_{n=1}^{\infty} a_n$. On occasion, it may prove to be more convenient to begin the summation from $n = 0$ or from $n = N$, where $N > 1$. When the initial integer in the summation is of no particular importance, we shall simply write $\sum a_n$.

Since we have defined series in terms of sequences, we already have many of the major results at hand in the theorems from the first part of this chapter. It is only a question of restating these theorems with our new terminology.

Theorem 12. (Cauchy's Criterion)

The series $\sum a_n$ is convergent if and only if, for every $\varepsilon > 0$, there is some integer N such that

$$|a_{m+1} + a_{m+2} + \ldots + a_n| < \varepsilon$$

for all $m, n \geq N$.

This is merely a restatement for infinite

series of theorem 9 of this chapter. In the theorem, the N depends on ε and, of course, the particular series.

One immediate consequence of this theorem is the necessary condition for the convergence of any series, namely $\lim_{n \to \infty} a_n = 0$. This condition is necessary but not sufficient for convergence. The classic example to show this is the harmonic series $\sum_{n=1}^{\infty} 1/n$. It is easy to see why this series fails to converge. The partial sums are not bounded. If we take $n = 2m$ in the Cauchy criterion, we have

$$|1/m+1 + 1/m+2 + \ldots + 1/2m| > |1/2m + \ldots + 1/2m|$$

$= m/2m = 1/2$, so that with $\varepsilon < 1/2$, the necessary condition for convergence is not satisfied.

The harmonic series, then, leads us to the conjecture that a necessary and sufficient condition that a series converges is that $\lim_{n \to \infty} a_n = 0$ and the partial sums s_n be bounded. But this conjecture is only partly correct. From the fact that every bounded monotonic sequence is convergent, we can immediately arrive at the following result.

Theorem 13.

Let $\{a_n\}$ be a sequence of nonnegative terms. Then the series, $\sum a_n$, generated by this sequence converges if and only if the partial sums s_n are bounded and $\lim_{n \to \infty} a_n = 0$.

On the other hand, if the series has both positive and negative terms, these two conditions, albeit necessary, are not sufficient for the convergence of the series. For consider the series

$$a_n = 1-1/2-1/2+1/4+1/4+1/4+1/4-1/8 \ldots -1/8+1/16+\ldots$$
$$\leftarrow 8 \text{ terms } \rightarrow$$

The pattern should be clear. It is evident that $\lim_{n \to \infty} a_n = 0$ and $s_n \leq 1$ for all n. However $\lim_{n \to \infty} s_n$ does not exist and so this series fails to converge.

If $\sum a_n$ and $\sum b_n$ both converge, then $\sum(a_n + b_n)$ converges, but the converse is not true. Just consider the series $\sum(-1)^n$ and $\sum(-1)^{n+1}$. Again, if $\sum a_n$ converges and $\sum b_n$ diverges, then $\sum(a_n + b_n)$ diverges.

If c is any constant and a_n converges, say $\sum a_n = A$, then $\sum c a_n$ converges and $\sum c a_n = cA$. The multiplication of series is a bit more involved.

We will return to this question at the end of this chapter.

The associative law for addition holds for finite sums but more care is needed with infinite series.

Theorem 14.

Let $\sum a_n$ be a convergent series and let $\sum b_n$ be the series derived from $\sum a_n$ by inserting parentheses without changing the order of terms, that is, terms from $\sum a_n$ are grouped together to form b_n. Then $\sum b_n$ converges and to the same value as $\sum a_n$.

That it is necessary that $\sum a_n$ be convergent for the conclusion to hold is shown by the series $\sum (-1)^n$.

An even more striking illustration of the care that is needed with associativity for infinite series, however, is the following example. Let $a_n = (-1)^{n+1}(1+1/2^n)$. Then $\sum a_n = (1+1/2) - (1+1/4) + (1+1/8) - (1+1/16) + \ldots$. Now suppose that we group the terms together pairwise, that is $b_n = a_{2n-1} + a_{2n}$. We then find $\sum b_n = (1/2-1/4) + (1/8-1/16) + \ldots$. Now as we shall shortly see, this latter series is a geometric series and it does converge. But since $\lim_{n\to\infty} a_n \neq 0$ the original series cannot converge.

If we are given a series $\sum a_n$, there are two questions that can be asked. First, is the series convergent or divergent? Then, if the series is convergent, to what value does it converge? The second question is difficult, often impossible, to answer. But for the question of convergence or divergence of a series, there are many tests.

We close this section by observing that a finite number of terms of a series may be omitted or may be modified without affecting the convergence or divergence of the series. Of course, such modifications do affect the value to which the series converges.

2. Series of Positive Terms.

As an introduction to the tests for convergence, let us consider one of the most important of all series, the geometric series. This series is material either directly or indirectly, in many of the tests to follow. The series $\sum_{n=0}^{\infty} a^n$, where a is a constant, is called a geometric series. This series is convergent for $|a| < 1$ and is divergent for $|a| \geq 1$.

For $|a| < 1$, we have $\sum\limits_{n=0}^{\infty} a^n = 1/(1-a)$.

Theorem 15.

Let $\sum a_n$ and $\sum b_n$ be two given series, and let $|a_n| < b_n$. Then, if $\sum b_n$ is convergent, $\sum a_n$ is convergent. If the two series have only nonnegative terms and if $a_n \leq b_n$, then, if $\sum a_n$ is divergent, $\sum b_n$ is divergent.

A series such as $\sum b_n$, with $a_n \leq b_n$, is said to dominate the series $\sum a_n$. From our remark above concerning the modification of a finite number of terms of a series not affecting the convergence or divergence, it is clear that we could have required only that $a_n \leq b_n$ for all $n \geq N$, with N fixed, and the conclusion of the theorem would be unchanged.

In the first part of the theorem, the absolute value is essential. For example, let $a_n = -1$ for all n and let b_n be nonnegative terms of any convergent series. Then $a_n \leq b_n$ for all n but $\sum a_n$ does not converge. Again, in the second part of the theorem, both series must be with nonnegative terms as this example also shows.

Is there a comparison test between series with both positive and negative terms? The answer, in general, is no. For instance, consider the series $\sum a_n$, where

$$a_n = \begin{cases} 1/n & n \text{ odd} \\ -1/n(n-1) & n \text{ even} \end{cases},$$

and the series $\sum b_n$, where $b_n = (-1)^{n+1}/n$.

By considering partial sums, it is not difficult to show that the first series diverges whereas, as we shall soon see, the second series, the alternating harmonic series converges. What is to be noted in this example is that $|a_n| \le |b_n|$ for all n.

Once the reader has acquired a sufficient collection of convergent and divergent series, another useful comparison test is the following one.

Theorem 16.

Let $\sum a_n$ and $\sum b_n$ be two series of positive terms and suppose $\lim_{n\to\infty} a_n/b_n$ exists and is not zero. Then either both series are convergent or both series are divergent.

If the limit of the ratio is zero, however, no conclusion can be drawn. Both series may converge, both may diverge, or, indeed, one may converge and one diverge as with $a_n = 1/n^2$ and $b_n = 1/n$. ($\sum 1/n^2$ will be seen to be convergent below.)

Theorem 17. (The Integral Test)

Let f be a nonnegative, nonincreasing function defined on some interval $[N,\infty)$, where N is a fixed positive integer, and let $a_n = f(n)$ for $n \geq N$. Then the series $\sum_{n=N}^{\infty} a_n$ converges or diverges according as the improper integral

$$\int_N^\infty f(t)\ dt = \lim_{x\to\infty} \int_N^x f(t)\ dt$$

is convergent or divergent.

This test is of limited use but it is one of the appropriate tests for a number of important series. Using the integral test, you can show that the series $\sum 1/n^p$ converges for $p > 1$ and diverges for $p \leq 1$.

It is important that the function in the theorem be nonincreasing on the whole interval. Without this condition, it is possible to construct continuous functions f such that $f(n) = a_n$ and the series converges but the integral diverges (and vice versa).

There is another convergence test for series of

142.

positive terms which handles many of the series for which the integral test is appropriate. This test is more convenient since it will often reduce the task of studying a given series to the consideration of a geometric series.

Theorem 18. (Cauchy's Condensation Test)

Let $\{a_n\}$ be a monotonic decreasing sequence of nonnegative numbers and let $\sum a_n$ be the series generated by this sequence. Then the series $\sum_{n=1}^{\infty} a_n$ converges if and only if the series

$$\sum_{n=0}^{\infty} 2^n a_{2^n} = a_1 + 2a_2 + 4a_4 + 8a_8 + \ldots \quad \text{converges.}$$

The reader would do well to apply the condensation test to the series $\sum 1/n^p$.

Consider the following series, where the lower limits of the summations are to be chosen in such a way that each series has only nonnegative terms.

$$\sum \frac{1}{n \cdot \ln n} \quad ; \quad \sum \frac{1}{n \cdot \ln \ n \cdot \ln \ (\ln \ n)} \quad ;$$

(A)
$$\sum \frac{1}{n \cdot \ln \ n \cdot \ln \ (\ln \ n) \cdot (\ln \ \ln \ \ln \ n)} \quad ; \quad \ldots .$$

$$\sum \frac{1}{n \, (\ln n)^2} \quad ; \qquad \sum \frac{1}{n \, \ln n \, (\ln \ln n)^2} \quad ;$$

(B)

$$\sum \frac{1}{n \, \ln n \, (\ln \ln n) \, (\ln \ln \ln n)^2} \quad ; \quad \ldots.$$

The pattern established in (A) and (B) should be clear. Now using theorem 17 or theorem 18, it is not difficult to show that each of the series of (A) diverges and each of the series of (B) converges. What is to be noticed is the following: as we move from left to right, each of the series in (A) diverges "more and more slowly". Similarly, moving from left to right, each of the series in (B) converges "more and more slowly". To put this in another way, the corresponding terms of the corresponding series in (A) and (B) differ by less and less as we move from left to right in the family of series. It would, therefore, be reasonable to conjecture that there must be some type of line of demarcation, at least in some vague sense, between these convergent and divergent series. This conjecture, however, is false, In fact, we have the following somewhat surprising result.

Theorem 19.

1. Let $\sum a_n$ be a divergent series of positive

terms. Then $\sum(a_n/s_n)$ diverges, where $s_n = \sum\limits_{k=1}^{n} a_n$.

2. Let $\int a_n$ be a convergent series of positive terms. Then $\sum(a_n/\sqrt{r_n})$ converges, where $r_n = \sum\limits_{k=n}^{\infty} a_k$.

If we note that for a divergent series $s_n \to \infty$, and for a convergent series $r_n \to 0$, what the theorem says is that, given any divergent series, it is always possible to construct a series that diverges more "slowly". Similarly, given any convergent series, it is always possible to construct a series that converges more "slowly".

There is another type of comparison test for series with positive terms.

Theorem 20.

Let $\int a_n$ and $\int b_n$ be series of positive terms, and suppose that $b_{n+1}/b_n \le a_{n+1}/a_n$ for all $n \ge N$, where N is a fixed integer. Then if $\int a_n$ converges, $\int b_n$ converges. If $\int b_n$ diverges, $\int a_n$ diverges.

Taking $\int a_n$ to be a geometric series, the reader should be able to formulate a convenient ratio test for series of positive terms.

3. Absolute Convergence.

Definition.

The series $\sum a_n$ is said to be absolutely convergent if the series $\sum |a_n|$ converges. If the series $\sum a_n$ is convergent but not absolutely convergent, then $\sum a_n$ is said to be conditionally convergent.

Theorem 21.

Every absolutely convergent series is convergent.

This theorem follows immediately from Cauchy's criterion for convergence. However, the converse of the theorem is false. The alternating harmonic series $\sum (-1)^n/n$, is convergent, but taking the absolute value of each term, we get the harmonic series which we have shown to be divergent.

But it is true that if a series with all positive terms converges, it must converge absolutely. In a similar fashion, if all the terms of a series are negative, and if the series converges, it must converge absolutely. What if a series has only a finite number of terms with one sign, say negative, and all the other terms are positive? Again, if such a series is to converge at all, it must converge absolutely.

We are now going to present a number of tests for absolute convergence. In general, the result of these tests will be one of three conclusions:

i.   the series is absolutely convergent;

ii.   the series is not absolutely convergent;

iii. the test is indecisive.

The reader should be warned that "not absolutely convergent" does not, in general, imply divergence. A series that is not absolutely convergent can either diverge or be conditionally convergent. Of course if the only way that a series can converge is absolutely, "not absolutely convergent" does imply divergence. Similarly, if a test is indecisive for a particular series, we can draw no conclusion. Such a series may converge absolutely, converge conditionally, or diverge.

Theorem 22. (The Ratio Test)

Let $\sum a_n$ be a given series such that $a_n \neq 0$ for $n \geq N$, where N is a fixed integer.

1. If $\overline{\lim} \, |a_{n+1}/a_n| < 1$, the series is absolutely convergent.

2. If $\underline{\lim} \, |a_{n+1}/a_n| > 1$, the series is not absolutely convergent.

3.   If $\underline{\lim} \, |a_{n+1}/a_n| \leq 1 \leq \overline{\lim} \, |a_{n+1}/a_n|$, the test

is indecisive.

Theorem 23.   (The Root Test)

Let $\sum a_n$ be a given series.

1.   If $\overline{\lim} \, (|a_n|)^{1/n} < 1$, the series is absolutely

convergent.

2.   If $\overline{\lim} \, (|a_n|)^{1/n} > 1$, the series is not ab-

solutely convergent.

3.   If $\overline{\lim} \, (|a_n|)^{1/n} = 1$, the test is indecisive.

If we replaced the limit superior and limit

inferior by the limit in the previous two theorems,

the theorems would remain valid, but we would be re-

stricting the applicability of these tests since

there are series for which the appropriate limits

fail to exist.

Theorem 24.

Let $\sum a_n$ be a given series. Then

$$\underline{\lim} \, |a_{n+1}/a_n| \leq \underline{\lim} \, (|a_n|)^{1/n} \leq \overline{\lim} \, (|a_n|)^{1/n}$$

$$\leq \overline{\lim} \, |a_{n+1}/a_n|.$$

What this theorem implies is that whenever the ratio test shows that a series converges absolutely or diverges, the root test will lead to a like conclusion. But there are series for which the ratio test is inconclusive whereas the root test is not. In this sense, theorem 23 is stronger than theorem 22. The former, however, is often the more easy test to apply.

To see an example of this, consider the series $\sum a_n$, where

$$a_n = \begin{cases} (\tfrac{1}{2})^{n^2} & n \text{ odd} \\ (\tfrac{1}{2})^{n^3} & n \text{ even} \end{cases} \text{, so that}$$

$$\sum a_n = \tfrac{1}{2} + (\tfrac{1}{2})^8 + (\tfrac{1}{2})^9 + (\tfrac{1}{2})^{64} + (\tfrac{1}{2})^{25} + \dots .$$

In this case, the ratio test fails spectacularly in that $\overline{\lim} \, (|a_{n+1}/a_n|) = \infty$ and $\underline{\lim} \, (|a_{n+1}/a_n|) = 0$. However, for this series we have $\overline{\lim} \, (|a_n|)^{1/n} = 0$, and so this series converges absolutely.

Neither the ratio test nor the root test will necessarily lead to a conclusion for all series. For example, both tests are indecisive for the series $\sum 1/n^2$, a series which we have previously determined to be convergent by other methods.

Theorem 25. (Raabe's Test)

Let $\sum a_n$ be a given series such that $a_n \neq 0$ for $n \geq N$, where N is a fixed integer.

1. If $\underline{\lim} \, n(1 - |a_{n+1}/a_n|) > 1$, the series is absolutely convergent.

2. If $\overline{\lim} \, n(1 - |a_{n+1}/a_n|) < 1$, the series is not absolutely convergent.

3. If $\underline{\lim} \, n(1 - |a_{n+1}/a_n|) \leq 1 \leq \overline{\lim} \, n(1 - |a_{n+1}/a_n|)$, the test is indecisive.

This test will often work where the root and ratio tests fail, as with the series $\sum 1/n^2$. However, this test can itself be indecisive while the root test may give a positive result. The example given after theorem 24 will again suffice to show this.

Theorem 26. (Gauss' Test)

Suppose that $|a_{n+1}/a_n|$ can be written in the for

$$|a_{n+1}/a_n| \;=\; 1 - \alpha/n \;+\; \beta_n/n^\rho \;,$$

where $\rho > 1$, and $\{\beta_n\}$ is a bounded sequence. Then the series $\sum a_n$ is absolutely convergent if $\alpha > 1$ and is not absolutely convergent if $\alpha \leq 1$.

The advantage of Gauss' test over Raabe's test is that the former test is not indecisive. It makes a definite statement, -- either the series is absolutely convergent or it is not, On the other hand, it may be difficult to put the ratio into the prescribed form.

For example, consider the series

$$\sum \frac{(a+1)(a+2) \ldots (a+n)}{(b+1)(b+2) \ldots (b+n)} \quad ,$$

where b is not to be a negative integer, so that all the terms in the series are well defined. Depending on the values of a and b, this series can have both positive and negative terms but eventually, all the terms will be of the same sign. Hence, if this series converges at all, it must converge absolutely. Now Raabe's test shows that the series converges for $b > a+1$ and diverges for $b < a+1$. What of the case $b = a+1$? Raabe's test is indecisive, but by Gauss' test we get divergence.

4. Rearrangement of Terms.

Consider a given series $\sum a_n$ generated by a sequence $\{a_n\}$. As we have indicated, if the series (and hence, the sequence) has only a finite number of terms of a given sign, absolute convergence or divergence is the only possibility. Let us assume, then, that the sequence $\{a_n\}$ has an infinite number of both positive and negative terms. We form two subsequences, $\{p_n\}$ and $\{q_n\}$, where $\{p_n\}$ is the subsequence of $\{a_n\}$ consisting of the positive terms of the original sequence and $\{q_n\}$ is the subsequence consisting of the absolute values of the negative terms.

For example, take $\{a_n\} = \{(-1)^{n+1}/n\} = \{1,-1/2,1/3,-1/4,\ldots\}$. In this case we have $\{p_n\} = \{1/(2n-1)\} = \{1,1/3,1/5,\ldots\}$ and $\{q_n\} = \{1/2n\} = \{1/2,1/4,1/6,\ldots\}$.

Using the notation above, we have the following theorem, the second part of which the reader may find rather surprising.

Theorem 27.

If the series $\sum a_n$ is absolutely convergent, then each of the series $\sum p_n$ and $\sum q_n$ is convergent. Moreover, $\sum a_n = \sum p_n - \sum q_n$. If, on the other hand, the series is conditionally convergent, then each of the series $\sum p_n$ and $\sum q_n$ diverges.

It is this divergent characteristic of both positive and negative parts of a conditionally convergent series that gives rise to Riemann's remarkable theorem on the rearrangement of terms in a series.

Definition.

A sequence $\{b_n\}$ is said to be a **rearrangement** of a sequence $\{a_n\}$ if there is a bijection $f: N \to N$ from the natural numbers onto the natural numbers such that $b_n = a_{f(n)}$ for each $n \in N$. A series $\sum b_n$ is said to be a rearrangement of a series $\sum a_n$ if $\{b_n\}$ is a rearrangement of the sequence $\{a_n\}$.

Now if $\sum a_n$ is an absolutely convergent series, and if $\sum b_n$ is a rearrangement of $\sum a_n$, it is not difficult to see that $\sum b_n$ converges and $\sum a_n = \sum b_n$. Moreover, it is a consequence of Riemann's rearrangement theorem that if $\sum a_n = \sum b_n$ for all rearrangements, $\sum b_n$, of $\sum a_n$, then $\sum a_n$ converges absolutely. The situation is entirely different for conditionally convergent series.

Theorem 28. (Riemann's Rearrangement Theorem)

Let $\sum a_n$ be a conditionally convergent series and let $\rho \in R$ be an arbitrary real number. Then there exists a rearrangement function f_ρ such that

153.

$\sum a_{f_\rho(n)} = \rho$. Moreover, there are also rearrangement functions f_∞ and $f_{-\infty}$ such that $\sum a_{f_\infty(n)} = \infty$ and $\sum a_{f_{-\infty}(n)} = -\infty$.

Thus a rearrangement of terms of a conditionally convergent series can change the value of the series, or even bring it about that the originally convergent series diverges. The reason this can occur is, of course, the fact that both positive and negative parts of a conditionally convergent series diverge.

The reader might find it amusing to test this theorem on one of the most simple conditionally convergent series, the alternating harmonic series,
$$\sum_{n=1}^{\infty} (-1)^{n+1}/n = \ln 2.$$

If the reader attempts to prove or consults a proof of theorem 28, he will see that complete freedom must be permitted in the rearrangement.

Suppose, on the other hand, that the rearrangement is bounded? By this we mean that no term of the original series is moved more than N places from its original position (N is a fixed integer). In this case, the rearranged series will converge and have the same value as the original series. It would be a worthwhile exercise for the reader to show this.

5. Convergence Tests for Arbitrary Series.

So far all our tests for convergence have been tests for series with nonnegative terms or tests for absolute convergence. What about a general series? Theorem 27 provides a limited test.

For example, the series $1-1/4+1/3-1/16+1/5-1/36 + \ldots$, that is the series $\sum a_n$, where

$$a_n = \begin{cases} 1/n & n \text{ odd} \\ 1/n^2 & n \text{ even} \end{cases} ,$$

must certainly diverge by that theorem. It is not absolutely convergent since the sum of positive terms, as a type of harmonic series, diverges. And it is not conditionally convergent since the series of even terms converges.

To test the convergence of a general series is often rather difficult, but there are two standard theorems which are useful.

Theorem 29. (Dirichlet's Test)

Let $\sum b_n$ be a series whose sequence of partial sums is bounded, and let $\{a_n\}$ be a nonincreasing sequence with $a_n \to 0$. Then the series $\sum a_n b_n$ converges.

Theorem 30.   (Abel's Test)

Let $\sum b_n$ be a convergent series and let $\{a_n\}$
be a nonincreasing bounded sequence. Then the series
$\sum a_n b_n$ converges.

The difference in the two tests should be noted.
In Dirichlet's test less is required of the series.
The sequence of partial sums need only be bounded;
it need not be convergent as in Abel's test. On the
other hand, Dirichlet's test requires more of the
sequence. The sequence must be nonincreasing and
converge to zero, whereas Abel's test only requires
that the sequence be nonincreasing and bounded.

For a highly nontrivial application of
Dirichlet's test, the reader might consider the
series
$\sum a_n$ = 1/2ln2 + 1/3ln3 - 1/4ln4 - ...- 1/7ln7 + 1/8ln8
+ ...,   where the signs appear in groupings of
2, 4, 8, 16, etc. It is clear that the sequence
$\{a_n\}$ = $\{1/\ln n\}$ monotonically decreases to zero.
The series, $\sum b_n$= 1/2 + 1/3 - 1/4 - ... - 1/7+ 1/8+...
in fact, diverges. However, the sequence of partial
sums is, indeed, bounded. Hence, by Dirichlet's test,
the original series is convergent.

Let the reader who attempts to prove that this sequence of partial sums is bounded be forewarned that he is undertaking a rather difficult problem.

There is one test for convergence which is actually a corollary of Dirichlet's test. However, it is important enough, albeit specialized, to be stated as a theorem in its own right. This is the alternating series test, where by an alternating series we mean one where the plus and minus signs appear successively. The general term in the series, therefore, will have a factor such as $(-1)^n$.

Theorem 31. (Leibnitz's Test)

Let $\sum a_n$ be a series whose terms are alternately positive and negative. Suppose that

1. $|a_{n+1}| \leq |a_n|$,

2. $a_n \to 0$ as $n \to \infty$.

Then $\sum a_n$ is convergent.

From this theorem, the convergence of the alternating harmonic series emerges immediately. However, for the test to be applicable, the series must be strictly alternating. For example, the series

$\sum a_n = 1/2 + 1/3 - 1/4 - 1/5 - 1/6 - 1/7 + 1/8 + \ldots,$

where the signs appear in groups of 2, 4, 8, 16, etc. satisfies the two conditions of the theorem, but is not an alternating series. In fact, this series fails to converge since it does not satisfy the Cauchy criterion, even though the sequence of partial sums is bounded, as we indicated above.

Both conditions of the theorem must be satisfied. For instance, the series $\sum(-1)^n$ is alternating and satisfies condition 1. But $a_n \not\to 0$ and, as we know, this series diverges. Or to use another example we have seen earlier, take $\sum(-1)^n a_n$, where

$$a_n = \begin{cases} 1/n^2 & n \text{ odd} \\ 1/n & n \text{ even} \end{cases} .$$

This alternating series satisfies the second condition of the theorem but not the first. It too fails to converge.

However, the theorem gives only a sufficient condition for the convergence of an alternating series. That the condition is not necessary is illustrated by the following example. Let

$$\sum a_n = \sum(-1)^{n+1} |\cos n| /n^2.$$

This alternating series does not satisfy the first condition of the test, but it is easily seen to be convergent. In fact, this series is absolutely convergent.

6. Multiplication of Series.

We have postponed until now the final topic we shall treat in this chapter, the multiplication of infinite series. Logically, perhaps, this subject should have appeared earlier but, as we shall see, it has subtilities which make a prior experience with series all the more useful for its consideration.

As the reader should have come to expect, the situation is quite satisfactory if we take the product of two absolutely convergent series.

Theorem 32.

Let the series $\sum_{n=0}^{\infty} a_n$ and $\sum_{n=0}^{\infty} b_n$ each be absolutely convergent, converging to the values A and B respectively. Then

$$\sum_{n=0}^{\infty} c_n = \sum_{n=0}^{\infty} a_n \cdot \sum_{n=0}^{\infty} b_n$$

is absolutely convergent with the value $A \cdot B$. Moreover, the absolute convergence of $\sum c_n$ is independent of the way in which the product is taken and the order in which the terms appear, as long as each term $a_i b_j$ in the product appears only once.

There are any number of ways by which we may take the product of two series. One natural way might be to take $c_0 = a_0 b_0$; $c_1 = a_0 b_1$; $c_2 = a_1 b_0$; $c_3 = a_1 b_1$. Then c_4, c_5, c_6, c_7, and c_8 would be $a_0 b_2$, $a_1 b_2$, $a_2 b_2$, $a_2 b_1$, and $a_2 b_0$ respectively. At the n th stage we would take all the terms $a_i b_j$ for which i and j are not greater than n, and for which at least one of them is equal to n.

There is a more convenient way of taking the product which is called Cauchy multiplication. With this method of multiplication we would have $c_0 = a_0 b_0$; $c_1 = a_1 b_0 + a_0 b_1$; $c_2 = a_2 b_0 + a_1 b_1 + a_0 b_2$ and so on. At the n th stage, c_n is the sum of those $a_i b_j$ such that i+j = n. Hence, we have for the Cauchy product

$$\sum_{n=0}^{\infty} c_n = \sum_{n=0}^{\infty} a_n \cdot \sum_{n=0}^{\infty} b_n = \sum_{n=0}^{\infty} \sum_{k=0}^{n} a_k b_{n-k}$$

If both of the series in the product are not absolutely convergent, more care is necessary.

Theorem 33. (Mertens)

Let $\sum_{n=0}^{\infty} a_n$ be an absolutely convergent series with sum A, and let $\sum_{n=0}^{\infty} b_n$ be a convergent series (not necessarily an absolutely convergent series) with sum B. Then the Cauchy product of $\sum a_n$ and $\sum b_n$ is convergent and $\sum_{n=0}^{\infty} c_n = A \cdot B$, where

$$c_n = \sum_{k=0}^{n} a_k b_{n-k} \cdot$$

So the Cauchy product of an absolutely convergent series and a conditionally convergent series converges. But the theorem does not say that the product converges absolutely. For instance, let

$$\sum a_n = \sum_{n=0}^{\infty} (-1)^n / (n+1)^{3/2} \text{ and } \sum b_n = \sum_{n=0}^{\infty} (-1)^n / (n+1)^{1/2}.$$

Then the Cauchy product is

$$\sum_{n=0}^{\infty} c_n = \sum_{n=0}^{\infty} ((-1)^n \sum_{k=0}^{n} 1/(k+1)^{3/2} \cdot 1/(n+1-k)^{1/2}).$$

A simple calculation shows that $|c_n| \geq 1/n$ so that, although the Cauchy product converges here, it does not converge absolutely.

What of the Cauchy product of two conditionally convergent series? In such a case, it may well be true that the product diverges. The example usually given to show this is the product of

$$\sum a_n = \sum_{n=0}^{\infty} (-1)^n/(n+1)^{1/2}$$ with itself. The reader

should verify that the Cauchy product does, indeed, diverge in this case.

But the Cauchy product of two conditionally convergent series need not diverge. In order to illustrate this, take $\sum a_n = \sum b_n = \sum_{n=0}^{\infty} (-1)^n/(n+1)$. In this case the Cauchy product does converge conditionally, as the reader may show.

However, perhaps the most surprising fact about the Cauchy product of two series is that it is possible for the Cauchy product of two divergent series to be convergent -- even absolutely convergent! For $n \geq 1$, let us take $a_n = a^n$ and $b_n = b^n$, where a and b are real numbers with $a \neq b$. If we choose a and b to satisfy the equations $a = (1 - b_0)(a - b)$ and $b = (a_0 - 1)(a - b)$, then $c_n = 0$ for $n \geq 1$. Now if we take $a - b = 1$, then $a_0 = b + 1 = a$ and $b_0 = 1 - a = -b$. Hence, with a and b integers which are such that $a - b = 1$, we will have

$$\sum a_n = a + a + a^2 + a^3 + \ldots \quad \text{and}$$

$$\sum b_n = -b + b + b^2 + b^3 + \ldots$$

Both series clearly diverge. However, the Cauchy product is $\sum\limits_{n=0}^{\infty} c_n = -ab + 0 + 0 + 0 + \ldots = -ab$, an absolutely convergent series. (For details on this example, we refer the reader to Gelbaum and Olmsted's "Counterexamples in Analysis", p. 62).

What, finally, can be said of the product of two convergent series? If the series converge only con-ditionally, we must, as illustrated above, treat each problem individually. The Cauchy product may be con-vergent or divergent. However, if the Cauchy product does converge, we have the following result.

Theorem 34.

Let the series $\sum a_n$ and $\sum b_n$ converge to A and B respectively. If the Cauchy product $\sum c_n$ of $\sum a_n$ and $\sum b_n$ converges to the value C, then

$$C = A \cdot B.$$

The most comprehensive reference to all the material of this and the following chapter is:
K. Knopp. Theory and Applications of Infinite Series. Hafner Publishing Company: New York. The interested reader would do well to consult this book.

VII. SEQUENCES AND SERIES OF FUNCTIONS.

1. Boundedness and Convergence.

In the previous chapter where we discussed numerical sequences and series, we started with a discussion of sequences and only after we had completed the discussion of sequences did we proceed to the discussion of infinite series. Our procedure in this chapter shall be somewhat different. We shall be treating the principal results on series and sequences of functions in the same sections. The major theorems in each section should indicate our reason for making such a choice.

When dealing with sequences of functions, the definition of sequence remains unchanged from the definition given in chapter 6. The only difference is that now the n th member of the sequence $\{f_n\}$ will be a function f_n defined on some domain which will generally be an interval. We will assume that all the functions in the sequence have D as their common domain of definition. For a fixed x_o in D, the sequence $\{f_n(x_o)\}$ is, of course, a numerical sequence in the sense of chapter 6. However, it will often be convenient to use the notation $\{f_n(x)\}$

for the sequence of functions. The context should generally make it clear whether $\{f_n(x)\}$ is to be considered to be a sequence of functions or a numerical sequence (x fixed). Similar remarks hold for series of functions.

In order to have some uniformity of notation, we shall write sequences as $\{f_n\}$, $\{g_n\}$, and $\{h_n\}$ and series as $\sum u_n$, $\sum v_n$, and $\sum w_n$, but we shall use f, g, and h for the functions represented by the series, that is $f(x) = \sum_{n=1}^{\infty} u_n(x)$.

We start by defining convergence for a sequence of functions.

Definition.

Let $\{f_n\}$ be a sequence of functions defined on some set D, and suppose for every x in D, the numerical sequence $\{f_n(x)\}$ converges to f(x). Then the sequence $\{f_n\}$ converges to the function f on D, that is, $f(x) = \lim_{n\to\infty} f_n(x)$ for all x in D.

In other words, given any $\varepsilon > 0$ and any x in D, there is an integer N such that $|f_n(x) - f(x)| < \varepsilon$ for all $n \geq N$.

What is to be noted is that the particular

integer N that satisfies the conditions of the definition depends both on ε and on the point x. In general, we cannot choose a single N such that the inequality holds for all x in D. A sequence of functions that converges in the sense of the definition above is called pointwise convergent.

We will illustrate this definition by three sequences which appear to be rather similar. Consider the sequences of functions $\{f_n\}$, $\{g_n\}$, and $\{h_n\}$, all defined on $[0,1]$ by

$$f_n(x) = n^2 \cdot x^n \cdot (1-x),$$

$$g_n(x) = n \cdot x^n \cdot (1-x),$$

$$h_n(x) = n^{\frac{1}{2}} \cdot x^n \cdot (1-x) .$$

It would be well for the reader to sketch a rough set of curves for each of these sequences.

Now $f_n(1) = g_n(1) = h_n(1) = 0$, and a simple application of l'Hôpital's rule shows that

$$\lim_{n \to \infty} n^p x^n = 0 \text{ for } 0 \le x < 1 \text{ and } p = 1/2, 1, 2.$$

Hence, for all x in the interval $[0,1]$, we have

$$f(x) = \lim_{n \to \infty} f_n(x) = 0,$$

$$g(x) = \lim_{n \to \infty} g_n(x) = 0,$$

$$h(x) = \lim_{n \to \infty} h_n(x) = 0.$$

Each of these sequences is a sequence of continuous functions, and each converges to a continuous function on [0,1], namely the constant zero function. Still, there are striking differences in each of these sequences. For example, each of the functions f_n, g_n, and h_n peaks (takes on a maximum) at precisely the same point, $x = n/(n+1)$. We find that

$$f_{n\ max} = n(n/n+1)^{n+1} \quad \xrightarrow{n \to \infty} n/e \rightarrow \infty :$$

$$g_{n\ max} = (n/n+1)^{n+1} \quad \xrightarrow{n \to \infty} 1/e \quad ;$$

$$h_{n\ max} = 1/n^{\frac{1}{2}}(n/n+1)^{n+1} \xrightarrow{n \to \infty} 1/n^{\frac{1}{2}}e \rightarrow 0.$$

Thus, for some point in the interval, some element of the sequence $\{f_n\}$ takes on an arbitrarily large value. Nevertheless, the sequence $\{f_n\}$ converges pointwise to zero on [0,1], as do the sequences $\{g_n\}$ and $\{h_n\}$.

Again, the behavior of $\{f_n\}$ and $\{g_n\}$ at their maximum shows that we could never expect to find an

integer N, independent of x, such that $|f_n(x)| < \varepsilon$
and $|g_n(x)| < \varepsilon$ for $n \geq N$ and for all x in [0,1]. On
the other hand $|h_n(x)| \leq |1/n^{\frac{1}{2}}e|$, so that if we
choose $N = (1/e\varepsilon)^2$, we will certainly have $|h_n(x)| <$
for all $n > N$ and for all x in [0,1].

The above examples, to which we shall return,
lead us to define another type of convergence.

Definition.

A sequence of functions $\{f_n\}$ defined on some
domain D converges uniformly to the function f if,
given any $\varepsilon > 0$, there is an integer N such that
$|f_n(x) - f(x)| < \varepsilon$ for all $n \geq N$ and for all x in D.

The distinction between pointwise convergence
and uniform convergence, then, is to be found in how
the integer N may be chosen. In the case of pointwise convergence, the integer N depends both on ε
and on the point x. We should, perhaps, write
$N = N(\varepsilon, x)$. In the case of uniform convergence, the
integer N depends on ε but is independent of the
point x in D. We could write, in this case, $N = N(\varepsilon)$

The reader should be saying, "I've seen some-
thing like this before." Indeed he has -- in the
distinction between continuity of a function on its

domain and uniform continuity. In the usual ε-δ definition, δ depends both on ε and the point x, that is $\delta = \delta(\varepsilon,x)$ for simple continuity, whereas uniform continuity has it that one δ suffices for all x in D, $\delta = \delta(\varepsilon)$.

From the definitions given, it should be clear that if a sequence of functions is uniformly convergent, it is pointwise convergent, but the converse is not true. The examples we have given above illustrate this. The three sequences $\{f_n\}$, $\{g_n\}$, and $\{h_n\}$ are all pointwise convergent on [0,1], but only $\{h_n\}$ is uniformly convergent on the interval.

We have given the definition of a Cauchy sequence for numerical sequences in chapter 6. We may similarly define a Cauchy sequence for a sequence of functions.

Definition.

1. The sequence of functions $\{f_n\}$ defined on D is a Cauchy sequence if, given any $\varepsilon > 0$ and given x in D, there exists an integer $N = N(\varepsilon,x)$ such that $|f_n(x) - f_m(x)| < \varepsilon$ for all $m,n \geq N$.

2. The sequence of functions $\{f_n\}$ defined on D is a uniformly Cauchy sequence if, given any $\varepsilon > 0$,

there exists an integer $N = N(\varepsilon)$ such that for $m, n \geq N$, $|f_n(x) - f_m(x)| < \varepsilon$ for all x in D.

We will leave it to the reader to show that a necessary and sufficient condition for the sequence $\{f_n\}$ to be uniformly convergent on D is that it be a uniformly Cauchy sequence on D.

It should be unnecessary to go into detail in order to define pointwise convergence and uniform convergence for a series of functions on some domain D. It is merely a question of applying the definitions above to the sequence of partial sums of the series. And it is a simple matter to formulate the Cauchy criterion for pointwise convergence and uniform convergence of a series of functions.

Before moving on to consider the rather important theorems of this chapter, we must discuss the idea of boundedness for a sequence and series of functions.

Definition.

A sequence of functions $\{f_n\}$ defined on a domain D is said to be pointwise bounded (or bounded) on D if, for each x_o in D, the numerical sequence $\{f_n(x_o)\}$ is bounded.

An example would be the sequence of functions $\{1/nx\}$ defined on the interval $(0,1]$. Given any x_0 in $(0,1]$, the sequence $\{1/nx_0\}$ is a bounded numerical sequence. The three sequences we considered at the start of this chapter are bounded sequences. On the other hand, the sequence $\{x^n\}$ defined on $[1,2]$ would not be a bounded sequence, even though each of the functions x^n is bounded on the interval.

The reader should carefully distinguish between a sequence of bounded functions and a bounded sequence of functions. Each of the functions $f_n(x) = 1/nx$ is unbounded on the interval $(0,1]$, but $\{1/nx\}$ is a bounded sequence. This example shows that a sequence of unbounded functions can converge to a bounded function on D.

Now a careful consideration of the definition of pointwise boundedness for sequences shows that, if for each x_0 in D the numerical sequence $\{f_n(x_0)\}$ is bounded, then we must have $|f_n(x_0)| < M$ for $n = 1,2,3,\ldots$ But the bound M varies from point to point in D. Perhaps, then, it would be better to write $M(x_0)$. With these remarks, and with a look back at the sequences $\{f_n\}$, $\{g_n\}$, and $\{h_n\}$ which were considered at the start of this chapter, the reader should immediately see what our next definition

171.

is going to be.

Definition.

A sequence of functions $\{f_n\}$ defined on D is said to be uniformly bounded on D if there is a positive number M such that $|f_n(x)| < M$ for all n = 1,2,3,... and for all x in D.

In this case, M is entirely independent of the x in D. The sequences $\{g_n\}$ and $\{h_n\}$ are examples of uniformly bounded sequences, whereas the sequence $\{f_n\}$ is not uniformly bounded.

The concepts of continuity, convergence, and boundedness are, of course, all different. But the analogy between continuity -- uniform continuity, convergence -- uniform convergence, and boundedness-- uniform boundedness should be clear to the reader.

The principal theorem connecting convergence and boundedness is the following one.

Theorem 1.

Let $\{f_n\}$ be a uniformly convergent sequence o bounded functions on D. Then $\{f_n\}$ is uniformly bounded on D.

The theorem gives a sufficient condition for t uniform boundedness of the sequence on D. But to

172.

draw the conclusion of uniform boundedness, both conditions must be present. For example, consider the sequence $\{f_n\}$ defined by

$$f_n(x) = \begin{cases} n & 0 < x \le 1/n \\ 1/x & 1/n < x \le 1 \end{cases} .$$

The f_n are all bounded functions on $(0,1]$, and, indeed, the sequence $\{f_n\}$ converges to $1/x$ on $(0,1]$. But this sequence is not uniformly convergent. In this example, the sequence is not uniformly bounded. In fact, we have here an example of a sequence of bounded functions which converges to an unbounded function.

On the other hand, consider the sequence $\{g_n\} = \{(1/x + 1/n)\}$ on $(0,1]$. It is easy to see that this sequence converges uniformly to $1/x$ on the interval, but the functions $g_n(x)$ are not bounded on the interval. Hence, $\{g_n\}$ can hardly be a uniformly bounded sequence. This example also illustrates that a sequence of functions on D need not be uniformly bounded in order that the sequence converge uniformly.

It is not necessary, however, that a convergent sequence of bounded functions be uniformly convergent in order that the sequence be uniformly bounded. A

sequence we saw earlier in this chapter, namely
$\{g_n(x)\} = \{nx^n(1-x)\}$ on $[0,1]$ gives one such illustration.

Theorem 1 does give a useful negative criterion
for the uniform convergence of a sequence of functions. For it follows from the theorem that if a sequence of bounded functions is not a uniformly bounded sequence, then the only way the sequence can converge is pointwise. Of course, the sequence need not
converge at all.

Theorem 2.

1. Let $\{f_n\}$ and $\{g_n\}$ converge uniformly on
D to f and g respectively. Then $\{(f_n + g_n)\}$ converges uniformly to $(f + g)$ on D.

2. Let $\{f_n\}$ and $\{g_n\}$ be uniformly convergent
sequences of bounded functions on D, and suppose
these sequences converge uniformly on D to f and g
respectively. Then the sequence $\{f_n g_n\}$ converges
uniformly to fg on D.

In part 2 of the theorem, the condition that
the members of the sequence be bounded functions
cannot be omitted. We can see this by using as an
example a sequence we have already used. Let

174.

$\{f_n\} = \{g_n\} = \{(1/x + 1/n)\}$ on $(0,1]$. Both sequences converge uniformly to $1/x$ on the interval, but $\{f_n g_n\}$ is not uniformly convergent on the interval. However, $\{f_n g_n\}$ does converge pointwise to $1/x^2$ on $(0,1]$.

Up to this point we have only considered sequences of functions, but similar results hold for series of functions if we consider the series to be defined in terms of the sequence of its partial sums. For example, we saw that a sequence of functions unbounded on an interval can converge uniformly. In a similar fashion, it is not necessary that each term of a series of functions be bounded in order that the series be uniformly convergent. Consider the series $f(x) = \sum_{n=-1}^{\infty} x^n = 1/x + 1 + x + x + x^2 + \ldots$ defined on the interval $(0,a]$, where $a < 1$. It is not immediately obvious that this series is uniformly convergent on the interval, but if the reader will recall the geometric series, he will see that this series sums to $1/x(1-x)$. From this it is easy to prove that the series is uniformly convergent.

2. Tests for Uniform Convergence.

It is usually relatively easy to test a sequence of functions for uniform convergence by returning to

175.

the definition of uniform convergence. However, un-like the example given above, the uniform convergence of a series of functions generally poses a more for-midable problem. In this section we wish to review the principal tests for the uniform convergence of a series.

The first test and, perhaps, the easiest to apply is a test which is an immediate consequence of the Cauchy criterion for convergence, the Weierstrass M-test.

Theorem 2. (Weierstrass M-test)

Let $\sum u_n(x)$ be a series of functions defined on D, and suppose there exists a convergent numerical series of nonnegative terms, $\sum M_n$, such that

$|u_n(x)| \leq M_n$ for all x in D and n = 1, 2, 3, ...

Then $\sum u_n(x)$ is uniformly convergent on D.

For example, consider the Riemann ζ-function which is defined by $\zeta(x) = \sum 1/n^x$. Does this series converge? We have seen that $\sum 1/n^p$ converges for $p > 1$ and diverges for $p \leq 1$. Hence, using this as our comparison series, we conclude that $\sum 1/n^x$ con-verges uniformly for all $x \geq 1 + \varepsilon$, where $\varepsilon > 0$.

The reader should note that the conditions given in the theorem also allow us to conclude that $\sum u_n(x)$ is absolutely convergent. Now there are series that are uniformly convergent but not absolutely convergent as we shall soon see. For such series the M-test is of no use. To deal with such series, we have Dirichlet's test and Abel's test. The reader should not neglect to compare these tests with the similar tests having the same name that were stated in chapter 6.

Theorem 5. (Dirichlet's Test)

Let $\{u_n(x)\}$ and $\{v_n(x)\}$ be sequences of functions defined on a common domain D, and suppose the sequence of partial sums $s_n(x) = \sum_{k=1}^{n} u_k(x)$ is uniformly bounded on D, that is $|s_n(x)| \leq M$ for all x in D and n = 1, 2, 3, ... Moreover, suppose $\{v_n(x)\}$ is a sequence converging uniformly to zero on D, the convergence being monotonic for every fixed x in D. Then the series $\sum u_n(x) \cdot v_n(x)$ converges uniformly on D.

Theorem 4. (Abel's Test)

Let $\{u_n(x)\}$ and $\{v_n(x)\}$ be sequences of functions defined on a common domain D, and suppose $\sum u_n(x)$ is uniformly convergent on D. Moreover,

177.

suppose $\{v_n(x)\}$ is a uniformly bounded sequence which is nonincreasing on D. Then $\sum u_n(x) \cdot v_n(x)$ is uniformly convergent on D.

Often enough in practice, one of the sequences $\{u_n(x)\}$ or $\{v_n(x)\}$ will, in fact, be a numerical sequence, so that the uniform convergence of $\{v_n(x)\}$ or $\sum u_n(x)$ will be automatic. The reader should be aware of the fact that a numerical series or sequence, if it converges at all, must converge uniformly. We shall leave the formulation of the alternating series test for uniform convergence as an exercise. It follows immediately from theorem 3.

As an example of the application of Abel's test, consider the series $\sum_{n=1}^{\infty} \dfrac{\ln(1 + nx)}{n\,x^n}$ defined on an interval $[a,\infty)$, where $a > 1$. Here we take $\{u_n(x)\} = \{1/x^{n-1}\}$ and $\{v_n(x)\} = \{1/nx\,\ln(1+nx)\}$. The M-test, with the comparison series the geometric series $\sum(1/a)^n$ shows that $\sum 1/x^{n-1}$ is uniformly convergent on the interval. The sequence $\{1/nx\,\ln(1+nx)\}$ is uniformly bounded on the interval. In fact, this sequence converges monotonically to zero. Certainly $1/nx\,\ln(1\_nx) \leq 1$ for all n and for all x in $[a,\infty)$. Thus, $\sum \dfrac{\ln(1 + nx)}{nx^n}$ is uniformly

convergent on $[a,\infty)$.

As an example of an application of Dirichlet's test, consider the series $\sum(-1)^n/(n+x^2)$, where the series is to be defined on the entire real line. Here we take $\{u_n(x)\} = \{(-1)^n\}$ and $v_n(x) = \{1/(n+x^2)\}$. Certainly, the partial sums $\sum\limits_{k=1}^{n} (-1)^k$ are uniformly bounded on R, and the sequence $\{1/(n+x^2)\}$ is uniformly convergent to zero, the convergence being monotonic for each fixed x. Hence $\sum(-1)^n/(n+x^2)$ is uniformly convergent on R.

A comparison with the harmonic series easily shows that this series is not absolutely convergent. Therefore, uniform convergence does not imply absolute convergence. We shall see examples to show that the implication does not go in the other direction either. Absolute convergence does not imply uniform convergence.

This example manifests another characteristic of uniformly convergent series of functions. The fact that $\sum u_n(x)$ converges uniformly in a region D does not imply that $\sum |u_n(x)|$ converges uniformly on D.

But if we consider the Cauchy criterion for uniform convergence, it is not difficult to see that the uniform convergence of $\sum |u_n(x)|$ on D does, indeed, imply the uniform convergence of $\sum u_n(x)$ on D.

Before we close this section, we wish to emphasize that the tests for uniform convergence give suffi-cient conditions for the uniform convergence. The conditions are by no means necessary.

3. Convergence and Continuity.

We have seen the connection between convergence and boundedness of a sequence or series of functions Now we look at the connection between convergence and continuity.

For example, if the reader will review some of the illustrations of section 1 of this chapter, he will see that a pointwise convergent sequence of continuous bounded functions may converge to an unbounded continuous function. However, if the con-vergence is uniform, this may not occur, although a sequence of continuous unbounded functions may con-verge uniformly to an unbounded continuous function.

It is a rather remarkable fact that a sequence of everywhere discontinuous functions may converge, indeed, converge uniformly, to a continuous function. One example to show this is rather elementary. Consider the sequence $\{f_n(x)\}$ defined on the interval $[0,1]$ -- we could just as well define the sequence on the entire real line -- by

$$f_n(x) = \begin{cases} 1/n & x \text{ rational} \\ 0 & x \text{ irrational} \end{cases}.$$

This sequence converges uniformly to the constant zero function.

Now our principal question in this section shall be as follows: when does a convergent sequence of continuous functions converge to a continuous function? For the limit of a convergent sequence of continuous functions need not be continuous. The most simple illustration of this fact is the sequence $\{f_n(x)\} = \{x^n\}$ defined on $[0,1]$. This sequence converges to the discontinuous function

$$f(x) = \begin{cases} 0 & 0 \le x < 1 \\ 1 & x = 1 \end{cases}.$$

Or to give an example of a series of continuous functions which does not converge to a continuous function, consider the series $\sum_{n=0}^{\infty} x^2/(1+x^2)^n$, defined on the entire real line. Each term is clearly continuous but, (think back to the geometric series), the series converges to the discontinuous function

$$f(x) = \begin{cases} 1 + x^2 & x \neq 0 \\ 0 & x = 0 \end{cases}.$$

What, then, is the sufficient condition that a convergent sequence or series of continuous functions converges to a continuous function?

Theorem 5.

Let the sequence of functions $\{f_n(x)\}$ be defined on an interval I, and suppose that the sequence converges uniformly to the function f on this interval. If each of the functions f_n is continuous at x_o in I, then f is continuous at x_o. If each f_n is continuous on I, then f is continuous on I.

The reader may formulate a similar theorem for a uniformly convergent series of functions. We exmphasize, however, that the theorem only gives a sufficient condition that the limit function be continuous. To see that uniform convergence is not a necessary condition, consider the sequence of continuous functions $\{f_n(x)\} = \dfrac{x^2}{x^2 + (1-nx)^2}$ on $[0,1]$. It is not difficult to see that this sequence converges to zero on the interval, but the convergence

s not uniform (consider the values of f_n at x= 1/n).
he first two examples of this chapter offer similar
llustrations.

Another angle from which to approach the question
f uniform convergence. is that of commutivity. Given
wo operations, when can we exchange the order of the
perations? For instance, given a sequence of func-
ions to be integrated, is the integral of the limit
f the sequence equal to the limit of the integrals?
r with the differentiable functions, is the limit of
he derivatives equal to the derivative of the limit?
e will answer these questions in the following sec-
ion.

With a sequence of continuous functions, the ques-
ion takes the following form. When do we have
$\lim_{\to a} \lim_{n\to\infty} f_n(x) = \lim_{n\to\infty} \lim_{x\to a} f_n(x)$? The sequence $\{x^n\}$ on
0,1] shows that equality is not always true. Uniform
onvergence of the sequence gives a sufficient con-
ition to change the order of the two limits.

Now we have seen that a uniformly convergent
equence of continuous functions converges to a con-
inuous function but the converse is, in general,
alse. However, under somewhat restrictive conditions,
e can assert a converse to theorem 5.

Theorem 6. (Dini)

Let $\{f_n(x)\}$ be a sequence of continuous functions defined on a closed bounded interval I. Suppose
that $f_n(x) \geq f_{n+1}(x)$ for all x in I, n = 1,2,3,...,
and that $\{f_n\}$ converges to a continuous function f
on I. Then $\{f_n\}$ converges uniformly to f on I.

The conditions in the theorem must all be present before we can conclude that the convergence is
uniform. For example, the sequence $\{x^n\}$ on [0,1]
is a monotonic sequence of continuous functions on
a closed bounded interval. But this sequence does
not converge to a continuous function. The convergence, in this case, is not uniform. If we consider
this same sequence on the interval [0,1), the limit
is, in this case, continuous, but the interval is no
closed. Again the convergence is not uniform. (We
might remark here that this sequence does converge
uniformly to zero on the interval [0,a], where a < 1.

The example immediately after theorem 5 shows
that monotonicity cannot be omitted in the theorem.

Finally, the sequence $\{f_n(x)\}$ defined on [0,1]

by

$$f_n(x) = \begin{cases} 0 & x=0 \text{ or } 1/n \le x \le 1 \\ -nx+1 & 0 < x < 1/n \end{cases}$$

is monotonic on the closed interval $[0,1]$, and converges to the continuous zero function. But the convergence is not uniform. In this case, you note that the f_n are discontinuous at the origin.

We are now in a position to take up an example we alluded to in chapter 3, namely an everywhere continuous nowhere differentiable function. At one time much thought was given to proving that the class of continuous functions and the class of continuous functions differentiable at at least one point were one and the same. The proof was not forthcoming, however, since these classes are, in fact, not the same. Weierstrass was able to show this by producing a function that is everywhere continuous but nowhere differentiable. Today many such functions are known. The one we give for an example is as "simple" as any but, of course, no such function can be expected to be altogether simple.

Let $\{x\}$ denote the distance from x to the nearest integer. For example, for $0 \le x \le 1$

$$\{x\} = \begin{cases} x & 0 \le x \le \tfrac{1}{2} \\ 1-x & \tfrac{1}{2} \le x \le 1 \end{cases} \quad ;$$

185.

$$\tfrac{1}{2}\,\{2x\} \;=\; \begin{cases} x & 0 \le x \le 1/4 \\[4pt] \tfrac{1}{2}-x & 1/4 \le x \le 1/2 \\[4pt] x-\tfrac{1}{2} & 1/2 \le x \le 3/4 \\[4pt] 1-x & 3/4 \le x \le 1 \end{cases}$$

The reader should plot $\tfrac{1}{4}\{4x\}$ on $[0,1]$ to make sure he has the picture of the behavior of such functions. One thing should be obvious. Such functions are continuous! We now define a sequence $\{f_n\}$ of continuous functions on $[0,1]$ by

$$f_n(x) \;=\; 1/10^n\,\{10^n x\} \quad.$$

We note that $|f_n(x)| \le 1/10^n$ for all x in $[0,1]$. If we wish, we may extend this sequence by periodicity to the entire real line, but this is not necessary for our purposes. Let a real valued function be defined on $[0,1]$ by

$$f(x) \;=\; \sum_{n=1}^{\infty} f_n(x) \;=\; \sum_{n=1}^{\infty} 1/10^n\,\{10^n\,x\} \quad.$$

The bound on each f_n shows that not only is this series of continuous functions convergent, it is uniformly convergent by the Weierstrass M-test. Hence, f is continuous on $[0,1]$. Is **f** differentiable? The answer is no. This function is not differentiable at any point in $[0,1]$.

We will not give the details of this, but the interested reader may consult Spivak's _Calculus_, p. 422, where the full details are given.

We would expect that functions with such a seemingly unusual property as to be nowhere differentiable yet everywhere continuous would be the exception rather than the rule. But this is not the case! In fact, if we consider the class of functions continuous on the unit interval, $C([0,1])$ and divide this class into two disjoint sets, say A and B, where A consists of those functions which are differentiable at at least one point and B consists of those functions which are nowhere differentiable on the interval, then B is "richer" than the set A. It would take us beyond the scope of this book to explain just what "richer" means in this context. For the reader with some knowledge of metric spaces, let it suffice to say that the set B is dense in $C([0,1])$ with the metric of uniform convergence.

4. Convergence and Integration
   Convergence and Differentiation

We return to the question of commutivity. That is, given a sequence of functions on [a,b] which converges to a function f, when do we have

$$\lim_{n \to \infty} \int_a^b f_n(x) \, dx = \int_a^b f(x) \, dx,$$

and $\quad \lim_{n \to \infty} f'_n(x) = f'(x)$?

The connection between convergence and integration is very similar to that between convergence and continuity. The connection is certainly considerably more simple than the connection between convergence and differentiability.

Let us reconsider the three sequences defined on [0,1] with which we started this chapter:

$$\{f_n(x)\} = \{n^2 x^n(1-x)\} \quad ;$$

$$\{g_n(x)\} = \{n \, x^n(1-x)\} \quad ;$$

$$\{h_n(x)\} = \{n^{\frac{1}{2}} x^n(1-x)\} \quad .$$

As we saw, all three sequences converge to the zero function, but only $\{h_n\}$ converges uniformly. Now

$$\int_0^1 f_n(x) \, dx = \frac{n^2}{(n+1)(n+2)} \quad \to \quad 1 \neq 0 \quad ;$$

$$\int_0^1 g_n(x) \, dx = \frac{n}{(n+1)(n+2)} \quad \to \quad 0 \quad ;$$

$$\int_0^1 h_n(x) \, dx = \frac{n^{\frac{1}{2}}}{(n+1)(n+2)} \quad \to \quad 0 \quad .$$

The first integral shows that pointwise convergence
of a sequence of functions is not sufficient to make
the limit of the integrals equal the integral of the
limit function. The second integral shows that uniform convergence of the sequence is not necessary in
order that the limit of the integrals equals the
integral of the limit function.

Our main result is as follows.

Theorem 7.

Let the sequence of functions $\{f_n\}$ be Riemann
integrable on the interval $[a,b]$, and suppose the
sequence converges uniformly to the limit function
f on $[a,b]$. Then f is Riemann integrable and

$$\lim_{n\to\infty} \int_a^b f_n(x)dx = \int_a^b \lim_{n\to\infty} f_n(x) \ dx \ (= \int_a^b f(x) \ dx).$$

The reader may formulate the similar theorem
appropriate for a uniformly convergent series of
functions.

Two remarks on this theorem are in order. First
note that the sequence in question is defined on a
bounded interval. We can see the point of this if
we consider the sequence $\{f_n\}$ defined for all
$x \geq 0$ by

$$f_n(x) = \begin{cases} 1/n & 0 \le x \le n \\ 0 & x > n \end{cases}$$

This sequence converges to $f(x) = 0$ on $x \ge 0$. And the convergence is uniform! But

$$0 = \int_0^\infty f(x)\,dx \ne \lim_{n \to \infty} \int_0^\infty f_n(x)\,dx = 1.$$

Again, the pointwise limit of a sequence of Riemann integrable functions need not be Riemann integrable. The standard example to show this is the following. Let $\{r_1, r_2, r_3, \dots\}$ be an enumeration of all the rational numbers on the unit interval $[0,1]$, and let $\{f_n\}$ be a sequence of functions defined on $[0,1]$ by

$$f_n(x) = \begin{cases} 0 & x \in \{r_1, r_2, r_3, \dots, r_n\} \\ 1 & x \in [0,1] - \{r_1, r_2, \dots, r_n\} \end{cases}.$$

Each f_n is Riemann integrable on $[0,1]$ since it is bounded and continuous at all but a finite number of points. However, it is not difficult to see that

$$f(x) = \lim_{n \to \infty} f_n(x) = \begin{cases} 0 & x \text{ rational} \\ 1 & x \text{ irrational} \end{cases},$$

a function that we have already seen to be not Riemann integrable on the unit interval. Here, of course, the sequence of functions fails to converge

190.

uniformly.

Does uniform convergence of a series or sequence of functions permit us to differentiate the
series or sequence term by term in order to obtain
the derivative of the limit function? Here the
answer must be negative. We have already seen a
spectacular counterexample in the everywhere continuous and nowhere differentiable function. In that
case the series of functions converges uniformly, or,
to put it in another way, the sequence of partial
sums converges uniformly. Each partial sum is differentiable at all but a finite number of points.
Nevertheless, the limit function is nowhere differentiable.

Or consider the sequence $\{f_n\}$ on $[-1,1]$ defined
by $f_n(x) = x/(1+n^2x^2)$. This sequence is uniformly
convergent to $f(x) = 0$ on the interval. Each term
is differentiable on the interval. Nevertheless,

$$0 = f'(x) \neq \lim_{n\to\infty} f'_n(x) = \begin{cases} 1 & x = 0 \\ 0 & 0 < |x| \leq 1 \end{cases}$$

The reader may verify that, in this case, the sequence of derivatives $\{f'_n\}$ is not uniformly convergent on the interval.

We may now formulate our principal theorem on convergence and differentiation.

Theorem 8.

Let $\{f_n\}$ be a sequence of functions defined and differentiable on the interval [a,b]. Suppose that

1. $\{f_n(x_0)\}$ converges for some point x_0 in [a,b],

2. $\{f'_n\}$ converges uniformly on [a,b].

Then $\{f_n\}$ converges uniformly on [a,b] to a differentiable function f, and

$$f'(x) = \lim_{n \to \infty} f'_n(x).$$

A similar theorem also holds for series of functions. In this case, we would require that the series of derivatives converges uniformly.

To give an illustration, consider the Riemann ζ-function which we have already seen to be defined by the series

$$\zeta(x) = \sum_{n=1}^{\infty} 1/n^x.$$

This series is uniformly convergent for $x \geq 1 + \epsilon$, where $\epsilon > 0$. Differentiating term by term, we obtain

192.

he series

$$- \sum_{n=1}^{\infty} \ln n / n^x \quad .$$

t is not difficult to show, using the M-test, that
this series itself is uniformly convergent for
$x \geq 1+\varepsilon$, $\varepsilon > 0$. Hence we conclude that the ζ-function
is differentiable for $x > 1$ with the derivative
given by

$$\zeta'(x) = - \sum_{n=1}^{\infty} \ln n / n^x .$$

We close this section by emphasizing once more
that, as with nearly all our theorems on uniform
convergence, theorem 8 gives only sufficient conditions for $f'(x)$ to be equal to $\lim f'_n(x)$. For
consider the sequence $\{f_n\}$ defined on the entire
real line by $f_n(x) = 1/n \cdot e^{-n^2 x^2}$. It is immediately
obvious that this sequence converges uniformly to
$f(x) = 0$ for all x. Again, it is fairly clear that
the sequence of derivatives $\{f'_n(x)\} = \{-2nx \; e^{-n^2 x^2}\}$
converges to zero. But in this case the convergence
is not uniform on any interval containing the origin.
Nevertheless, $f'(x) = \lim_{n \to \infty} f'_n(x)$ for all x. Hence,
uniform convergence of the sequence of derivatives
is not a necessary condition for the commutivity of
the limit operation with the taking of the derivative.

One final remark: condition 1 in the theorem, namely that $\{f_n\}$ should converge at at least one point, may seem rather strange. The reason for this is that the sequence of derivatives may converge, even uniformly, without the original sequence converging. The obvious example of this would be the sequence $\{f_n\}$ defined on all R by $f_n(x) = n$. In this case, $f'_n(x) = 0$ for all x. So $\{f'_n\}$ is uniformly convergent, whereas $\{f_n\}$ does not converge at any point.

5.   Power Series.

We now wish to discuss a particular type of series of functions, power series. A power series is a series of the form

$$\sum_{n=0}^{\infty} a_n (x - x_o)^n$$

where x_o is some fixed real number. By making a translation of the origin, this series takes on the form

$$\sum_{n=0}^{\infty} a_n x^n.$$

Since no loss of generality will occur under such a translation, we will restrict our attention to power series of the latter form.

Some very important functions are defined only in terms of their power series. But we can also start with a given function and ask what, if any, is its series representation in terms of power series. We shall discuss both questions. We begin with a consideration of the principal properties of power series.

In order to orient the discussion, let us consider a series of functions which is not a power series,

$$\sum_{n=1}^{\infty} \sin(nx)/n^2.$$

This series is a well behaved one, uniformly and absolutely convergent for all x. If we differentiate term by term, we obtain the series

$$\sum_{n=1}^{\infty} \cos(nx)/n.$$

This series converges for some values of x, for instance $x = \pi/2$, and diverges for other values of x, for example $x = 0$. If we differentiate term by term once more, we obtain the series

$$- \sum_{n=1}^{\infty} \sin(nx)$$

which diverges for all values of x except those for which x is an integral multiple of π. We shall see that a power series has no such aberrant behavior.

Consider now a power series expanded about the origin $\sum a_n x^n$. (For convenience, we shall omit the lower limit of summation except where it is essential the upper limit is to be understood to be ∞.) If this series is convergent at some point $x_o \neq 0$, it is not difficult to see that it is convergent, indeed absolutely convergent, for all $|x| < |x_o|$. But an even stronger result is possible.

Theorem 9.

Let the power series $\sum a_n x^n$ be convergent for $x = x_o \neq 0$, and suppose $|a_n x_o{}^n| < K$ for some constant K and for all n. Then $\sum a_n x^n$ is absolutely convergent for all $|x| < |x_o|$.

Now for a given power series there are only three possibilities. The series may converge absolutely for all values of x, as, for example $\sum_{n=0}^{\infty} x^n/n!$, the series representing the exponential function. The series may converge only for $x = 0$ as, for example $\sum n! \, x^n$. In this case the power series actually converges only in an improper sense. Finally, there is a positive number R such that $\sum a_n x^r$

converges absolutely for $|x| < R$ and diverges for $|x| > R$. This number R is called the radius of convergence and the interval $|x| < R$ is the interval of convergence of the power series. In case the series converges for all x, we say the series has an infinite radius of convergence.

As an aside, we note that R is called the "radius of convergence"; the more proper context within which we should discuss power series would be complex analysis. In this context, the interval of convergence becomes a **circle** of convergence.

A simple application of the Weierstrass M-test lets us determine the uniform convergence of any given power series.

Theorem 10.

Let the power series $\sum a_n x^n$ have a positive or infinite radius of convergence R. Let $0 < r < R$. Then $\sum a_n x^n$ converges uniformly to a continuous bounded function f for all $|x| \leq r$.

The reader will note that we have not said anything about the behavior of the series when $x = R$. There is good reason for this, as the consideration of the following three power series will show.

197 .

First consider the geometric series

$$\sum_{n=0}^{\infty} x^n.$$

In this case, as we know, R = 1. For x = 1 and x = -1 this series diverges. So this series converges absolutely for $|x| < 1$. The series

$$\sum_{n=1}^{\infty} x^n/n$$

again has a radius of convergence R = 1. In this case for x = 1 we obtain the divergent harmonic series, whereas for x = -1, we have the convergent alternating harmonic series.

Finally, consider

$$\sum_{n=1}^{\infty} x^n/n^2 .$$

Again R = 1. But here the series converges both uniformly and absolutely for $|x| \leq 1$.

Thus, in general, nothing can be said about the behavior of the power series at $|x| = R$. Each case must be examined individually.

The question of finding the radius of convergence of a power series now arises. A simple consideration of the ratio test for infinite series shows that the radius of convergence for the series $\sum a_n x^n$ is given by

$$R = \lim_{n\to\infty} |a_n/a_{n+1}|$$

provided this limit exists or is ∞. This is how we found the radius of convergence for the series above. The problem is that this limit does not always exist.

For example, consider the power series $\sum_n a_n x^n$ where

$$a_n = \begin{cases} 2^n & n \text{ odd} \\ 3^n & n \text{ even} \end{cases}.$$

In this case we have $|a_n/a_{n+1}| = 1/3(2/3)^n$ for n odd and $1/2(3/2)^n$ for n even. Hence, $\lim |a_n/a_{n+1}|$ does not exist. We do have $\underline{\lim} |a_n/a_{n+1}| = 0$ and $\overline{\lim} |a_n/a_{n+1}| = \infty$. And it is clear that $\underline{\lim} |a_n/a_{n+1}| \leq R \leq \overline{\lim} |a_n/a_{n+1}|$, but this is of no help.

Cauchy's root test gives us a more reliable method for obtaining the radius of convergence of a power series.

Theorem 11.

The radius of convergence of a power series is

given by

$$R = \frac{1}{\overline{\lim_{n \to \infty}} |a_n|^{1/n}} \qquad .$$

Thus, in the above example $R = 1/3$. The reader should distinguish carefully between

$1/\overline{\lim} |a_n|^{1/n}$ and $\overline{\lim} 1/|a_n|^{1/n}$. The example above shows that these are different.

In order to obtain the radius of convergence, it is usually easier to calculate $\lim |a_n/a_{n+1}|$, butif this limit does not exist or if the coefficients $|a_n|$ are such that their n th root is a relatively simple expression, it is best to resort to theorem 11.

In theorem 10 we saw that a power series converges to a continuous bounded function in any closed interval properly contained in the interval of convergence of the series. In fact, the series actually converges to a continuous function on the entire open interval of convergence. The following theorem is a consequence of the fact that a uniformly convergent series may be integrated term by term.

Theorem 12.

Let the power series $\sum\limits_{n=0}^{\infty} a_n x^n$, with radius of convergence R, define a function f on its interval of convergence. Let α and β ($\alpha < \beta$) be points in this interval. Then

$$\int_{\alpha}^{\beta} f(x)\ dx = \sum_{n=0}^{\infty} a_n/n+1\ (\beta^{n+1} - \alpha^{n+1}).$$

The theorem says nothing about the case where α or β is equal to R. Similarly, theorem 10 says nothing about the continuity of the function defined by the power series at x = R. There is a theorem due to Abel that enables us to consider these cases.

Theorem 13. (Abel)

Suppose the series $\sum a_n x^n$ has a radius of convergence R and that the series converges for x = R. Then the series converges uniformly on the closed interval [0,R].

A similar result holds, of course, if the series converges at x = -R. An immediate consequence of the theorem is that the function f, defined by the power series, is continuous on [0,R], and

$$\lim_{x \to R^-} f(x) = \lim_{x \to R^-} \sum a_n x^n = \sum a_n R^n.$$

The converse of Abel's theorem, however, is not true. For example, consider the series $\sum_{n=0}^{\infty} (-1)^n x^n$ which has radius of convergence $R = 1$.

In this case, $\lim_{x \to 1^-} \sum (-1)^n x^n = \lim_{x \to 1^-} 1/(1+x) = 1/2$. However, $\sum (-1)^n$ does not converge.

Abel's theorem enables us to extend the limits of integration on a function defined by a power series to R (or -R) provided that the series obtained by integrating term by term is convergent at $x = R$. That is $\int_0^R f(x) \, dx = \sum a_n/n+1 \ R^{n+1}$ provided the right hand side converges. It does not matter how the series $\sum a_n x^n$ behaves at $x = R$. However, if $\sum a_n x^n$ does not converge at $x = R$, the integral may have to be interpreted as an improper integral.

It is by means of Abel's theorem that the last theorem of chapter 6 is proven.

Theorem 14.

Let $f(x) = \sum a_n x^n$ and suppose that the power series has a radius of convergence R. Then the series $\sum n a_n x^{n-1}$ has the same radius of convergence and $f'(x) = \sum n a_n x^{n-1}$.

As a consequence of this important theorem, if f is defined by a power series with radius of convergence R, then f is infinitely differentiable and each of the power series representing the derivatives of f has a radius of convergence R. The reader should be struck by the contrast between this behavior of a power series and the behavior of the series obtained by term by term differentiation of $\sum \sin nx/n^2$ with which we began this section.

Power series can also be divided. The important point here is that the leading term of the series in the denominator does not vanish.

Theorem **15.**

Let $f(x) = \sum a_n x^n$ and $g(x) = \sum b_n x^n$ have radii of convergence R_1 and R_2 respectively. Assume that $b_o \neq 0$. Then there exists a positive number $\rho \neq 0$ such that $f(x)/g(x)$ has a power series expansion for $|x| < \rho$

where $0 < \rho \leq R = \min(R_1, R_2)$.

It is somewhat tedious to find the coefficients c_i in terms of the a_i and b_i, where $f(x)/g(x) = \sum c_n x^n$. This can be done, however, by means of long division or by using the method of undetermined coefficients. By the latter we mean equating the coefficients of like powers of x on both sides of the equation

$$\sum_{n=0}^{\infty} a_n x^n = \sum_{n=0}^{\infty} b_n x^n \sum_{n=0}^{\infty} c_n x^n = \sum_{n=0}^{\infty} x^n \sum_{k=0}^{n} b_k c_{n-k} \cdot$$

This method may seem somewhat inefficient, but if the reader uses it to obtain the power series for tan x = sin x/cos x, and then attempts to obtain the power series for tan x by any other method, he will see that the division of power series has its advantages. This is especially true if only the first few terms of the expansion are needed.

We have already considered Taylor's formula with a remainder,

$$f(x) = \sum_{k=0}^{n} f^{(k)}(x_0)/k! \ (x-x_0)^k + R_{n+1} \cdot$$

Suppose, now, that the function f has derivatives of all orders in some open interval containing the point $x = x_0$. Moreover, let us suppose that the remainder terms $\{R_n\}$ form a null sequence, that is $\lim_{n \to \infty} R_n = 0$. In this event, we have

$$f(x) = \sum_{n=0}^{\infty} f^{(n)}(x_o)/n! \; (x - x_o)^n.$$

This is, of course, a power series for the function f expanded about the point $x = x_o$. If $x_o = 0$, the resulting series is generally called the Maclaurin series. Hence, if we start with a given function, and wish to represent that function as a power series $\sum a_n x^n$, we see what relationship exists between the coefficients a_n and the function itself, $a_n = f^{(n)}(0)/n!$ Moreover, if the function is defined by the power series $\sum a_n(x-x_o)^n$, then we have $a_n = f^{(n)}(x_o)/n!$

Theorem 16.

If two power series $\sum a_n x^n$ and $\sum b_n x^n$ are equal in the interval $|x| < R$ on which both series converge, then the two series are identical, that is $a_n = b_n$ for all n.

In other words, the power series expansion for the function is its Taylor series. Thus, the power series expansion for a function is unique.

6. Analytic Functions

We will end this chapter by considering briefly a special class of functions, real analytic functions.

Definition.

A function f is analytic on an open interval I if for each x_0 in I, there exists a power series $\sum a_n(x-x_0)^n$, and a positive number R such that for all $|x - x_0| < R$, $f(x) = \sum a_n(x - x_0)^n$.

It is to be understood, of course, that the open neighborhood of x_0, $|x - x_0| < R$, is contained in I.

Not every function is analytic. For example, it follows from theorem 14 that if f is analytic at x_0, it is infinitely differentiable on the interval $|x - x_0| < R$. Hence, $f(x) = |x|$ is not an analytic function at the origin. From the uniqueness of the power series expansion, it follows that the power series in the definition must be the Taylor series of the function.

A function f is analytic on an open interval

I if for each point x_o in I, the function is represented by its Taylor series in some subinterval of
I centered at x_o.

Now it is easy enough for the reader to construct nonanalytic functions, but the reader may be
surprised to learn that even if a function is infinitely differentiable at a point, it need not be
analytic there. The classic example of this is
the function

$$f(x) = \begin{cases} 0 & x = 0 \\ \\ e^{-1/x^2} & x \neq 0 \end{cases} \quad \cdot$$

The reader may show that this function is infinitely
differentiable at the origin, and, using l'Hôpital's
rule, see that $f^{(n)}(0) = 0$ for all n.   Hence, the
Taylor series for f gives $f(x) = 0 + 0 + \ldots$
for all points in some open interval containing
the origin. For the given function, this is patently
absurd.

Under what circumstances, then, may a function
be represented by its Taylor series on some open
interval?

Theorem 17.

Let f be defined and infinitely differentiable

on some open interval I. Then a necessary and
sufficient condition that f be analytic at a point
x_0 in I is that there exist positive numbers M and
ρ such that $|f^{(n)}(x)| \leq M^n$ n! for all n and for
all $|x - x_0| < \rho$.

It would be a worthwhile exercise for the
reader to show that the infinitely differentiable
nonanalytic function given above fails to satisfy
the conditions of the theorem.

The usual algebraic operations hold for analytic
functions.

Theorem 18 .

If f and g are analytic at some point x_0 of an
open interval I, then

1) f + g is analytic at x_0;
2) f$\cdot$g is analytic at x_0;
3) if $f(x_0) \neq 0$, 1/f is analytic at x_0.

We shall leave it to the reader to properly
formulate an appropriate theorem for the analyticity
of the composite function. It is analytic!

We end this section with a theorem which should
show that if the above material appears to be

straightforward, there are subtleties that may escape the less than careful reader.

Theorem 19.

Suppose that f is analytic at every point of the interval $|x - x_0| < R$. Moreover, suppose the Taylor series for f converges at every point of the same interval. Then the Taylor series converges to f at every point of the interval $|x - x_0| < R$.

We first note that there is a difference between the Taylor series converging and the Taylor series converging to f. The nonanalytic function discussed above is an illustration of this. But the main point the reader should note is that the theorem does not say that if the function f is analytic at every point $|x - x_0| < R$, then the Taylor series converges to f at every point of the same interval. In general, this will not be true.

A simple illustration of this would be the function $f(x) = 1/(1+x^2)$. It is a consequence of theorem 18(3) that f is analytic on the entire real line. Now the Taylor series of f expanded about the origin is $\sum_{n=0}^{\infty} (-1)^n x^{2n}$, a power series with radius of convergence 1.

The reader should not be surprised to be some-what mystified by this. The reason is that the proper context within which to discuss analytic functions is actually complex variables. A consideration of $f(z) = 1/(1+z^2)$ as z approaches $+\sqrt{-1}$ or $-\sqrt{-1}$ will show the source of the difficulty.

The reader may also consider the nonanalytic function discussed above in the complex plane. The behavior of e^{-1/z^2} in the neighborhood of z = 0 may shed some light on the fact that the Taylor series for this function represents the function only at the origin.